Photoelectrochemistry

Photoelectrochemistry

Yu. Ya. Gurevich, Yu. V. Pleskov, and Z. A. Rotenberg

Academy of Sciences of the USSR
Moscow, USSR

Translated from Russian by
Halina S. Wroblowa
University of Pennsylvania

Translation edited by
Halina S. Wroblowa and
B. E. Conway
University of Ottawa

CONSULTANTS BUREAU • NEW YORK AND LONDON

Library of Congress Cataloging in Publication Data

Gurevich, ÍÙriĭ ÍAkovlevich.
  Photoelectrochemistry.

  1. Photochemistry. 2. Photoelectricity. I. Pleskov, ÍÙriĭ Viktorovich, joint author.
II. Rotenberg, Zakhar Aronovich, joint author. III. Title.
QD715.G8713   541'.35   78-21541
ISBN-13: 978-1-4613-3932-8   e-ISBN-13: 978-1-4613-3930-4
DOI: 10.1007/978-1-4613-3930-4

The Russian text underlying this translation was prepared by the authors especially
for this edition. This translation is published under an agreement with the Copyright
Agency of the USSR (VAAP).

FOTOELEKTROKHIMIYA
Yu. Ya Gurevich, Yu V. Pleskov, and Z. A. Rotenberg

Translation Editors' Foreword

We enthusiastically welcome this opportunity to introduce this major work of Gurevich, Pleskov, and Rotenberg to English-speaking readers since photoelectrochemistry has, in recent years, become very significant for modern energy transfer and energy conversion phenomena. While having its roots in early electrochemistry, this field, in its modern aspects, has had an important impact on knowledge of the production and state of solvated electrons and on photoassisted electrolysis at semiconductors. Photoeffects resulting in electron emission into solution have also given rise to new ways of understanding double-layer structure and measuring potentials of zero charge. Electrochemical photoemission studies have added to and complemented the literature of solvated electron chemistry arising from experiments with high-energy radiation.

The authors' treatment of photoelectron emission phenomena at metal/solution interfaces is thorough and quantitative and, we believe, will constitute a landmark in the development of this fundamentally interesting and practically important area of electrochemistry and photophysics.

H. Wroblowa
B. E. Conway

Foreword

A characteristic feature of modern electrochemistry is the continually broadening utilization of nontraditional methods and development of new directions of research. A number of such approaches are based on illumination techniques. First, irradiation is used in electrochemistry mainly as a research tool. Mention should be made here of methods such as electro-reflection, ellipsometry, internal reflection spectroscopy, interferometry of surface layers, and other techniques firmly established in experimental electrochemistry. Second, light directly affects electrode processes. Investigation of the latter phenomenon is the subject of photoelectrochemistry.

Among various directions of modern photoelectrochemistry, one of the most interesting and promising is that connected with the electron photo-emission effect at the electrode–solution interface. Interest in photoemission phenomena appeared among electrochemists as early as the 1920s. However, photoemission became a subject of independent study in electrochemical systems only in the last decade. The modern stage of development of this field was initiated by the experimental work of Barker. Its further rapid development has been connected with the creation of a sufficiently complete quantitative theory of photoemission in solution. A number of areas of research in this field have been carried out in England, the USA, East Germany, Czechoslovakia, and Argentina. In the USSR, significant results were obtained in this field in the Institute of Electrochemistry of the Academy of Sciences of the USSR and in the Branch of the Institute of Chemical Physics of the Academy of Sciences of the USSR. The accumulated experimental and theoretical material has demonstrated that photoelectron emission at the electrode–solution interface is a new and rather successful tool for investigating the structure and properties of the interface and the physicochemical processes occurring in its vicinity.

The authors of this book have themselves made significant contributions to studies of photoemission of electrons in electrochemical systems.

This book is of high scientific value and describes the basic part of the presently existing material. At the same time, it is easily understood by the investigator to whom the problems discussed here were previously alien. The book will undoubtedly be of interest to scientists working in electrochemistry, physical chemistry, and radiation chemistry.

Academician A. N. Frumkin

Preface

Hundreds of papers have appeared within the last decade treating various aspects of the effects of light at the electrode–solution interface. This is about an order of magnitude more than during the last half-century. There are several reasons for this rapid increase of interest in the physicochemical effects of illumination: availability of new light sources; rapid development of new methods and spectacular successes attained by means of optical methods in solid state physics; important advances in radiation physics and radiation chemistry in studies of the solvated electron and realization of its importance for the solution of a number of physicochemical problems; various theoretical and experimental investigations of the behavior of excess epithermal electrons in condensed media; finally, the specific properties of electrochemical systems which enable the photoprocess under given conditions of external illumination to be relatively easily and effectively controlled. As a result, a whole new field of electrochemical research "came to light" which was totally obscure until recent times.

A large quantity of material has accumulated concerning photoelectron emission in solutions. This provides the basis for presentation in the form of a monograph.

In our previous book *Modern Photoelectrochemistry: Photoemission Phenomena*, co-authored with A. M. Brodskii and published by Nauka (Moscow), an attempt was made to generalize the existing results of theoretical and experimental studies of photoemission. During the two years which have passed since the Russian edition of the present monograph was published, tremendous progress has been made in the field. First, the range of systems studied has increased considerably. Whereas in the early years all work was practically limited to the mercury electrode, at present solid metals, semiconductors, and dielectrics are of primary interest. Second, the "center of gravity" has shifted from the study of photoemission itself to the solution of purely electrochemical and radiation-chemical problems. For example, the original photoemission methods were developed to investigate the fine structure of the electric double layer and the kinetics of chemical and electrochemical reactions. Utilization of these methods leads, in certain cases, to results which are unobtainable, or difficult to obtain, by traditional techniques. Finally, certain new mechanisms of electrode photoexcitation have been suggested, such as, for example, "hole emission," contribution of surface plasmons to photoemission, etc. Among other factors, rapid progress in the field has been due to the application of lasers and other modern light sources in photoelectrochemical experiments.

One of the aims of the present book, especially the American edition, is to stress the close ties between photoemission and problems of electrochemistry and radiation chemistry.

Apart from including new results, a major part of the material had to be rewritten. In particular, we have attempted to present the theory of photoemission in a more accessible form that does not require the reader to be acquainted with quantum mechanics. Moreover, the experimental material can be followed using the final theoretical formulas which are presented for convenience at the beginning of the respective experimental sections.

The authors wish to express their gratitude to the late Academician A. N. Frumkin. Without his friendly support and valuable advice, the book could not have been written. We are also indebted to L. I. Krishtalik for his valuable comments. We want to take advantage of this opportunity to mention the contribution made by A. M. Brodskii to the development of a number of theoretical problems in photoemission. We also thank V. V. Eletskii, M. D. Krotov, V. I. Lakomova, Yu. A. Prishchep, and S. V. Sheberstov, who co-authored a considerable part of the work described in this book.

Photoemission phenomena in electrochemistry deserve the steady attention of experimental electrochemists, physical chemists, radiation chemists, physicists, and technologists. If the publication of this book elicits a broader interest in photoelectrochemistry, the authors will consider their goal fulfilled.

Contents

Notation

A	proportionality coefficient in the 5/2 power law
A_{op}	photoionization energy
A_s	real solvation energy of an electron
c_e	concentration of solvated electrons
c_A	concentration of electron acceptors
c_{el}	concentration of the electrolyte
c'	concentration of electron capture products [eA]
c^{**}	concentration of an acceptor
d	thickness of the compact double layer
$\mathscr{D}_e$	diffusion coefficient of the solvated electron
$\mathscr{D}_A$	diffusion coefficient of the acceptor
$\mathscr{D}'$	diffusion coefficient of the capture product [eA]
e	absolute value of the electron charge ($= 4.8029 \cdot 10^{-10}$ cgse)
E_f, E_i	final and initial energies of the electron, respectively
E_m	maximum energy of emitted electrons
E_s	energy of a solvated electron (reorganization energy of the medium)
ε_x	the x component of the electric field intensity
F	Faraday number
$\hbar$	$h/2\pi$ ($= 1.0546 \cdot 10^{-27}$ erg-sec), $h =$ Planck's constant
i	$\sqrt{-1}$
I	photoemission current
I_e	reverse current of solvated electrons returning to the electrode
I_{eA}	current produced by [eA] products reacting at the electrode
$\tilde{I}$	photoemission current under conditions of oscillating illumination
j	measured photocurrent
j_h	warm-up current (experimentally measured photocurrent)
$\vec{j}_H, \overleftarrow{j}_H$	cathodic and anodic currents for discharge and ionization of hydrogen atoms
j_x	partial flux of emitted electrons
J	power of the light flux absorbed by the electrode
J_0	light intensity
k	Boltzmann's constant ($= 1.3804 \cdot 10^{-16}$ erg-deg^{-1})
k_A	rate constant of the interaction of a solvated electron with an acceptor
k_r	rate constant of the recombination reaction

k_s	rate constant of the process of solvated electron capture by the electrode surface
k_{eA}	rate constant of the surface reaction of [eA] products
k_{ad}	rate constant of adsorption
k_v	rate constant of the homogeneous reaction of [eA] products in solution
$\vec{k}_1, \overleftarrow{k}_1$	rate constants of the cathodic and anodic electrode reactions of nonadsorbed hydrogen atoms
$\vec{k}_2, \overleftarrow{k}_2$	the same for adsorbed hydrogen atoms
m_0	free electron mass ($= 9.1085 \cdot 10^{-28}$ g)
P	vector of the final momentum of the emitted electron (outside the emitter)
p	the x component of the vector of the emitted electron momentum
p_y, p_z	the y and z components of the momentum, respectively
$\mathbf{p}_\parallel$	$\{p_y, p_z\}$
p_i, p_f	vectors of the initial and final momentum (quasi-momentum) of the electron in a solid, respectively
Q	$= (k_A c_A / \mathscr{D}_e)^{1/2}$
Q_Ω	$= [(k_A c_A + i\Omega)/\mathscr{D}_e]^{1/2}$
Q_v	$= (k_v / \mathscr{D}')^{1/2}$
R	gas constant
T	absolute temperature
U_{sv}	work of electron transfer from vacuum into a delocalized state in solution
V_{ms}	Volta potential at the metal–solution interface
$V(x)$	potential energy of the electron outside the metal
V_e	applied external potential
$V_e{}^*$	threshold potential
w	work function
$w^{(th)}$	equilibrium (thermodynamic) work function
$w^{(ph)}$	photoelectric work function
w_{mv}	work function of a metal in vacuum
w_{ms}	work function of a metal in the electrolyte
w_{sv}	work function of an electron in solution
x	coordinate
x_0	mean distance (from the electrode) of formation of solvated electrons
Y	quantum yield
$\hat{Z}_{ph}$	photoimpedance of an electrochemical system
α, β	transfer coefficients of electrochemical reactions
δ	range of the surface forces
δ_n	experimental dispersion of the photocurrent measurement

δ_N	thickness of the Nernst diffusion layer
Δ	angle between the plane of incidence of the polarized light and the plane of electric vector oscillations
ε	dielectric constant
θ	surface coverage with adsorbate
ϑ	angle of incidence
Θ	angle of the phase shift between the photocurrent and light intensity
κ^{-1}	Debye length
λ	De Broglie wavelength of the emitted electron
μ	chemical potential of electrons
ν, ν_{eA}, ν_R	stoichiometric coefficients
φ	electrode potential
φ_0	threshold potential of photoemission
φ_k	potential measured under experimental conditions
φ_{zc}	potential of zero charge
φ_{ph}	photopotential
$\Phi(x)$	source function of solvated electrons
χ_m, χ_s	surface potential
χ	electron affinity of a semiconductor
ψ	wave function of the electron
ψ'	potential of the outer Helmholtz plane = potential of inner limit of diffuse layer
ψ_f	time-independent wavefunction of the electron in the final state
ψ_f^*	conjugate wavefunction to ψ_f
ω	light frequency
ω_0	threshold frequency
ω_p	plasma frequency
ω_s	frequency of the "surface plasmons"
ω^*	second photoelectric threshold
Ω	frequency of modulation of the light intensity

Introduction

0.1. Historical Background

The study of photoelectron emission is a new and, at the same time, a very old field of electrochemistry. It is new because the nature of photo-emission has been elucidated only recently, within the last 12–15 years. It is old because the first observations of the photoeffect in electrochemical systems based (as we understand it now) on photoemission date from the early part of the 19th century.

The history of investigations of so-called photovoltaic (photogalvanic) phenomena at the metal–electrolyte interface is over 130 years old. Becquerel (1839) was the first to observe the appearance of an electric current upon illumination of one of two identical electrodes immersed in dilute acids. The phenomenon, called the *Becquerel effect*, became subsequently the subject of detailed investigation. Numerous studies which followed the pioneer work of Becquerel were carried out on a variety of systems including both the metal–electrolyte and metal–vacuum interfaces. The most important work of the latter kind was that of Hertz (1887), which led to the discovery of photoelectron emission at the metal–vacuum interface.

Photoelectron emission consists in the escape of electrons from the surface of a body illuminated by electromagnetic radiation, the electrons being delocalized in their final state. If the latter situation obtains, e.g., there is delocalization in the molecule or a group of molecules located at atomic distances from the emitting surface, the phenomenon is called a photoelectrochemical reaction.

Photoelectron emission, which was investigated in detail somewhat later (1888a) by Stoletov, quickly attracted scientific interest. It has been intensively studied both theoretically and experimentally for many decades (e.g., by Brodskii and Gurevich, 1973; Dobretsov and Gomoyunova, 1966; Görlich, 1962; Hughes and Du Bridge, 1932). The results obtained were of extreme importance in the development of the quantum theory of light, in solid state physics, and in other problems of modern physics. Hertz's discovery strongly stimulated the production of numerous and variegated apparatus in photocell electronics.

The success accompanying the production of these devices, particularly of vacuum phototubes, had for a long time diverted attention from a wider investigation of electrochemical photocells. Nevertheless, a variety of photo-sensitive electrochemical systems have since been devised, and theories concerning the origin of the photosignal have been proposed. A classification of photoelectrochemical cells according to the mechanism of their operation can be found in the review by Copeland *et al.* (1942) covering the early papers, as well as in the more recent work of Berg *et al.* (1967). All photo-sensitive systems can be arbitrarily divided, according to the nature of the processes involved, into three groups.

The first group comprises systems in which the photocurrent originates because of absorption of light by the solution, resulting in homogeneous photochemical reactions. In this case, the change of the electrode potential (or current) is due to the formation of excited molecules, free radicals, and other photolytic products which can undergo reduction or oxidation at the electrode. The latter, however, do not directly participate in the photoprocess in this case but play the part of an "indicator" for photochemical reactions. Such systems (cf. the reviews by Airey, 1973, and Honda, 1969) will not be discussed here.

The second group of systems, in which the photosignal arises mainly as a consequence of a photoconduction effect (i.e., the appearance, in a solid, of current carriers which take part in an electrode reaction), also remains beyond the scope of this book. These phenomena involve, for example, semiconductor electrodes (Myamlin and Pleskov, 1967; Gerischer, 1966a), metals coated with oxide films (Young, 1961) or dyes (Hillson and Rideal, 1953), and insulators (Mehl and Hale, 1967).

The third group of photovoltaic cells involves systems consisting of a clean metal surface in contact with an electrolyte which does not absorb light. It was these simplest systems that were the object of Becquerel's studies.

The following observations were described by Becquerel in 1859. Illumination of metals immersed in an electrolyte results in the appearance of an electric current, the direction of which depends on the pH of the electrolyte. The violet part of the spectrum was found to be most effective. The magnitude of the photocurrent was also found to depend strongly on the electrode polarization. These seemingly simple observations led to an important conclusion: The magnitude of the photosignal depends on the characteristics of the illuminating light, the electrode potential, and the electrolyte composition. Owing to the absence of the required type of apparatus, Becquerel could not carry out, in his time, any quantitative measurements.

The Becquerel effect has, however, been widely investigated sub-sequently; among others, Sihvonen (1926) suggested possible mechanisms leading to the appearance of a photocurrent, including the light-induced transfer of an electron from the metal into solution (particularly in the form of a solvated particle).

Audubert (1923), who studied the Becquerel effect on platinum, copper, and mercury, pointed out the similarity of the mechanism of the appearance of the photosignal to that of photoemission at the metal–vacuum interface discovered by Hertz. Later, however, Audubert changed his opinion, suggesting that the current results from photolysis of water molecules adsorbed at the electrode surface (Audubert, 1930).

Clark and Garrett (1939) were the first to determine the threshold wavelength above which the Becquerel effect virtually disappears. It was approximately the same for silver, gold, and copper.

Anand and Bhatnagar (1928) studied the effect of light on the decomposition of alkali metal amalgams by water. The rate of this process was found to depend not only on electrode illumination but also on the state of polarization of light. A greater increase in the reaction rate is observed when the electric vector of the light wave is directed perpendicularly to the metal surface than for the parallel orientation of the vector. This observation seems to confirm the above-mentioned hypothesis of Sihvonen concerning the possibility of electron emission from the metal into solution.

The work of Bowden (1931), and later Hillson and Rideal (1949), concerning the effect of light on the rate of electrochemical evolution of hydrogen and oxygen deserves special attention. Hydrogen evolution was accompanied by relatively low photocurrents (of the order 10^{-5} A/cm^2). The logarithm of current was found to depend approximately linearly on the energy of the quantum; the threshold wavelength was also determined. A similar dependence was observed in the case of anodic oxygen evolution; however, the threshold wavelength was not evaluated.

Hillson and Rideal interpreted the observed effects assuming that the illuminating light activated a fraction of the hydrogen atoms adsorbed at the electrode surface, thus increasing their rate of recombination and electrochemical desorption. (This interpretation obviously pertains to metals with intermediate or high coverage with atomic hydrogen.) For anodic processes, the authors assumed that the light decomposes molecules of the surface oxide thus creating active oxygen atoms which, before deactivation, can react at the electrode surface.

The diversity of systems studied and of interpretations of observed photoprocesses clearly indicates that the Becquerel effect consists of various phenomena. Only recently has the effect been resolved into its "elementary components."

The increased interest in photoelectrochemistry in the early sixties was due mainly to the work of Berg, M. Heyrovsky, and Barker who independently proposed views on the nature of the photosignals arising at illuminated metal electrodes.

Berg studied the effect of light on the dropping mercury electrode using the well-known method of polarographic analysis (Berg is the originator of the word "photopolarography"). According to Berg, the photocurrent arises because absorption of light increases the energy of electrons in the bulk of the metal and thus increases the probability of electrode reactions proceeding on such a "hot" electrode (Berg, 1960a, b; 1961; 1966; 1968; Berg and Schweiss, 1965).

Heyrovsky (1965; 1966a, b; 1967a, b; 1973) has interpreted the photo-effect in terms of light-induced decomposition with charge transfer of a complex formed by the electrode metal with the solvent or solute molecules. The type of polarization of the chemical bond in such a complex depends on the electron-donating or electron-accepting character of the adsorbate. Absorption of light quantum results in rupture of this bond and the transfer of electrons either to the electrode (anodic photocurrent) or to the adsorbed particle (cathodic photocurrent). It should be mentioned that similar views on photodesorption and photodecomposition of surface compounds were expressed much earlier by Veselovskii (1946) who treated them as elementary events of photogalvanic processes.

Barker and Gardner (1965; cf. 1973a for a summary of earlier studies) demonstrated the important role played in the electrochemical photoeffect by photoelectron emission from the metal into solution, as well as by sub-sequent interactions of emitted electrons with the solvent and solutes. The idea of photoelectron emission from metals into solutions was, as shown above, not entirely new; however, Barker and his co-workers were the first to prove (not directly but by other sufficiently reliable methods) the reality of electron photoemission into solutions and to define clearly the limits of applicability of their concepts (Barker *et al.*, 1966). Their work stimulated further development of experimental and theoretical photoemission studies. Simultaneously with the increasing number of experiments, the quantum-mechanical theory of photoemission into electrolytes was developed (Gurevich *et al.*, 1967), and investigation of the phenomenon proceeded in various countries. At present, work of this type is carried out in the USA, USSR, Argentina, Japan, and Australia.

The three hypotheses discussed above are not mutually exclusive. One or another mechanism can be dominant, depending on the conditions of the given experiment. The theories of Berg and Heyrovsky will be discussed further in Chapter 10. It is to be noted that electrochemical systems bereft of any photoactive substances in the bulk of solution and at the electrode surface have a wide range of electrode potentials and light frequencies over which photoelectron emission plays the principal part in the total photo-process. Such systems are the main subject of this book.

0.2. Stages of Photoelectron Emission into Solution

Photoelectron emission from metals into vacuum consists only of two stages:

1. The photoemission proper, i.e.; the electron transfer through the interface, following the absorption of a light quantum. This stage, in its turn, in some cases can be considered as consisting of several consecutive processes in the emitter (see Section 1.1).
2. The movement of emitted electrons in vacuum under the influence of an external field, from the cathode-emitter to the anode-collector.

As opposed to the photoemission of electrons into vacuum, photo-emission into an electrolyte is only the first stage of a complex, multistep process invariably involving solvated electrons.

The existence of solvated electrons, e.g., in metal–ammoniacal solutions, has been known for a long time. However, up to relatively recent times, these particles were considered by the majority of workers to exist only in rather exotic solvents or under special experimental conditions. Within the last two decades, however, the solvated electron became, thanks to radiation chemistry, a commonly accepted reactant in chemical kinetics. In particular, the formation of a hydrated electron in water (first indicated as possible by Platzman, 1953) occurs in a number of radiation-chemical and photo-chemical reactions. Some qualitative properties and characteristics of the solvated electron will be discussed in Sections 4.4 and 9.6.

According to Barker *et al.* (1966) the overall photoprocess includes the following steps (Fig. 0.1):

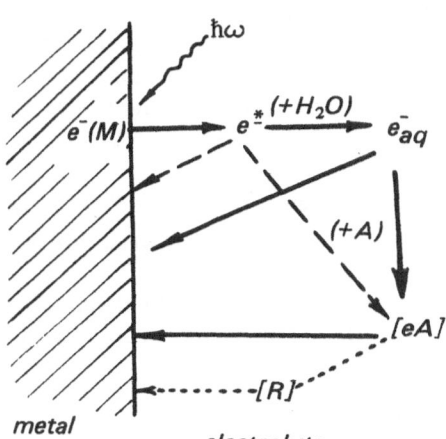

Fig. 0.1. Transitions of the emitted electron in aqueous solutions. $e^*_\pm$, emitted electron; e^-_{aq}, hydrated electron.

1. Photoemission itself.
2. Thermalization† and solvation (in aqueous solutions, hydration) of emitted electrons, resulting in a decrease of their initial energy to the mean energy of thermal movement and formation of the solvated electron e_s^- (in aqueous solutions, e_{aq}^-). The process is completed within ca. 10^{-12} sec (Pikaev, 1970; Mozumder and Magee, 1975).
3. Subsequent transformations in the solution of solvated electrons. The latter participate in chemical reactions with solution components (A) capable of capturing electrons; such components are called scavengers:

$$e_s^- + A \rightarrow [eA]$$

Several ions, e.g., H_3O^+, NO_3^-, NO_2^-, and molecules such as N_2O, O_2, and others can serve as scavengers.
4. Finally, the products of capture of solvated electrons by scavengers, [eA], formed in the vicinity of the electrode surface may participate in electrochemical and chemical reactions (the latter are indicated in Fig. 0.1 by the dashed line).

In the absence of scavengers in the solution, solvated electrons return to the electrode and the resulting stationary photocurrent is close to zero.

Moreover, very low currents (the so-called "residual photocurrents") (Barker and Gardner, 1965; Berg *et al.*, 1967) are observed even in the absence of specially introduced acceptors in the solution. They are usually connected with reactions of the solvated electron with the solvent and with traces of impurities (difficult to control in solutions) which act as scavengers (Barker and McKeown, 1975), or with recombination of hydrated electrons (Barker *et al.*, 1974b).

In principle, the possibility that nonsolvated (and even nonthermalized, i.e., "hot") electrons can participate in stage 3 should not be rejected *a priori*. This will be discussed in detail in Sections 2.1 and 5.2. The possible homogeneous reactions and the return to the electrode of nonsolvated electrons are shown in Fig. 0.1 by the dotted line.

0.3. General Features of Emission Phenomena in Electrochemical Systems

The qualitative properties of photoemission into electrolyte solutions which distinguish this process from emission into vacuum and dielectrics

---

† That is, when the emitted electron has attained thermal equilibrium with the surrounding solvent medium (Eds.).

are mainly connected with the existence of the electric double layer at the metal–solution interface (Frumkin *et al.*, 1952; Delahay, 1965). One side of the double-layer charge distribution is formed by the metal charge, the other by ions in solution. The ionic side can be divided in the simplest case into two parts: the compact and diffuse layers. The compact layer consists of ions closely attracted to the metal surface and extends usually only to atomic distances from the metal. The diffuse layer is formed by relatively free ions distributed at distances from the surface usually much larger than ionic radii. Ions in the diffuse layer are statistically distributed. The concentration of ions of opposite sign to that of the metal charge increases, and that of ions of the same sign decreases, with decreasing distance from the surface.

The potential drop between the metal and solution is distributed over the compact and diffuse layers. The magnitude of the potential drop in the diffuse part (the ψ' potential) essentially depends on the bulk concentration of the electrolyte. This problem is discussed in greater detail in Section 1.5. Here, we shall only mention that the relative contribution of the ψ' potential decreases with increasing concentration, becoming negligibly small for electrolyte concentrations of the order of 10^{-1} mole-liter^{-1} and higher. Thus, in sufficiently concentrated solutions, the electric field virtually disappears outside the compact layer.

The basic features of photoelectron emission into electrolytic solutions which differentiate the phenomenon from emission into vacuum are listed below.

1. The existence at the metal–solution interface of the electric double layer, in which virtually all the total externally applied potential drop occurs, results in the appearance of an additional parameter in photoemission, the electrode potential. Whereas in sufficiently concentrated electrolytic solutions the potential remains almost unchanged outside the compact layer (if the ohmic drop in solution is neglected), in metal–vacuum systems the potential drop extends over the whole space between the anode and cathode. If the electric field in the vicinity of the cathode is too low to induce field emission (which first becomes noticeable for fields of the order of 10^7 V-cm^{-1} (cf. Dobretsov and Gomoyunova, 1966)) the work function of the metal (in vacuum) is independent of the externally applied potential difference. The intensity of the electric field in the vicinity of the cathode in vacuum photoemission measurements does not usually exceed a few hundred V-cm^{-1} in order to avoid field emission and the so-called Schottky effect (i.e., the field-induced change of the barrier height at the metal–vacuum interface).

Conversely, the externally applied potential difference φ in the metal–electrolyte system results in a change of the energy level of the emitted electron in solution outside the double layer by $e\varphi$ (where e denotes the absolute value of the electronic charge). In other words, the electronic work function should change (Fig. 0.2) according to the relation

$$w_{ms}(\varphi) = w_{ms}(0) + e\varphi \qquad (0.1)$$

where $w_{ms}(0)$ is the work function of the electron transferred from metal into solution at the arbitrary zero potential of the electrode, and $w_{ms}(\varphi)$ is the work function at a potential φ.

Thus, the first essential difference in the behavior of metal–vacuum and metal–solution interfaces is that, in electrochemical systems, as opposed to metal–vacuum systems, the externally applied field results in a change of the electronic work function.

2. As opposed to the case of emission into vacuum, photoemission into the electrolyte transfers the electron into a condensed medium, bringing about, in general, an additional energy gain due to the interaction of the electron with the medium. Thus, under similar conditions (in particular, at uncharged interfaces) the electronic work function at the metal–solution interface differs (being usually lower) from that at the metal–vacuum interface.

3. Owing to the existence of the electric double layer, as well as of the condensed medium into which the electrons are emitted, the motion of the electron in the vicinity of the electrode differs from its motion in vacuum, where it experiences the action of an external field and of image forces.

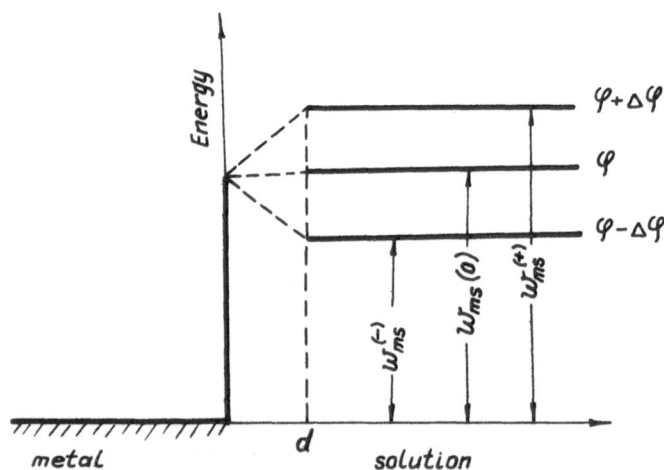

Fig. 0.2. Schematic diagram illustrating the relation between the electronic work function in solution and the electrode potential. w_{ms} (0), $w_{ms}^{(+)}$, and $w_{ms}^{(-)}$ are the work functions at φ, $\varphi + \Delta\varphi$, and $\varphi - \Delta\varphi$ potentials, respectively (where $\Delta\varphi > 0$); d is the thickness of the double layer.

The emitted electron interacts in solution with the solvent and solute molecules. The image forces in solution are to a great extent, or completely, screened by ions. All this is reflected in the quantum-mechanical description of the electron motion. It will be shown below how these properties lead to a different dependence of the photocurrent on the frequency of radiation from that observed for emission into vacuum.

4. Photoelectron emission into vacuum is a purely physical phenomenon, free of subsequent chemical reactions. Photoemission into electrolytes necessarily initiates chemical reactions in the solution (cf. Section 0.2). For example, electron transfer into the electrolyte results in reduction of some of its components.

It follows from the above discussion that electrochemical photoemission exhibits a number of properties which in a certain sense resemble common electrochemical reactions. This is reflected in several fundamental relationships.

The present state of the subject includes the formulation of ideas concerning photoemission in electrochemical systems, knowledge of the basic laws of the act of emission itself and of subsequent processes, and elucidation of the role played by the double layer in the photoemission method in electrochemical studies.

Fundamentals of the Theory of Photoelectron Emission from Metals into Solutions

1.1. Qualitative Description of the Phenomenon

The elementary act of photoemission originates from the interaction of photons with the metal, which results in energy transfer to the metal electrons. If the energy acquired by electrons is sufficiently high, they escape from the metal, giving rise to the photoemission current.

The rather complex mechanism of photoelectron emission should be qualitatively analyzed in two stages: starting with the microscopic properties of the interaction of incident light with the metal and then describing the microscopic process of the appearance of the photocurrent. In an analysis of the interaction of light with a metal, determined mainly by the behavior of conduction electrons, the range of light frequencies ω can be approximately divided into three intervals (cf., e.g., Ziman, 1972). The first interval includes frequencies $\omega < \tau^{-1}$, where τ is the relaxation time of excited states in the conduction band; in the normal temperature range, τ is of the order 10^{-13} sec. This interval has little interest with respect to photoemission: the energy of the quanta is such that electron transitions become energetically possible only upon absorption of a very large number of photons. The second frequency range is determined by the condition $\tau^{-1} < \omega < \omega_p$, where ω_p is the plasma frequency, usually of the order 10^{16} sec^{-1} (in the simplest model $\omega_p^2 = 4\pi e^2 n_e/m_0$, where n_e is the bulk concentration of conduction electrons and m_0 is the mass of the free electron). Usually, the second range contains the threshold frequency ω_0 (photoelectric threshold) given by $\hbar\omega_0 = w$, where $\hbar$ is Planck's constant and w is the electronic work function. (Strictly speaking, w is the photoelectric work function $w^{(ph)}$, which differs in the general sense from the thermodynamic work function $w^{(th)}$; this problem is discussed in more detail in Sections 4.3 and 9.1.) In this range, the reflection coefficient is close to unity. Finally, the third frequency range, $\omega > \omega_p$, corresponds to crossing the threshold of "ultraviolet transparency": The reflectivity of the metal as $\omega > \omega_p$ becomes very low so that electromagnetic waves penetrate deeply into the metal bulk. The photocurrent usually sharply increases in the vicinity of the threshold of ultra-violet transparency, i.e., at $\omega \simeq \omega_p$; the frequency $\omega_* \simeq \omega_p$, corresponding to the start of this sharp rise, is sometimes called the second photoelectric threshold (Dobretsov and Gomoyunova, 1966). The existence in metals of free electrons and a band structure does not qualitatively affect the general picture, although some changes can occur in the character of reflection and absorption, especially at frequencies corresponding to resonance transitions.

The modern theory of photoemission has its origin in the work of Einstein, who interpreted the existence of the photoelectric threshold and the independence of the final energy of emitted electrons on the intensity of radiation in terms of the quantum theory of light. However, his theory, which played a historic part in the development of ideas concerning the interaction of light with matter, proved incomplete in view of subsequent photoemission studies.

The complete theory of photoemission requires the solution of a complex quantum-statistical problem concerning interaction of electromagnetic waves with matter and including a number of collective effects, as well as properties due to the presence of boundaries in the system (cf., e.g., Feibelman and Eastman, 1974). Since it is hopeless to attempt a general solution of this problem, various approximate treatments of photoemission are used. The simplest model of the metal, still often used, is that of Sommerfeld. According to Sommerfeld, the metal consists of an electron gas enclosed in a box with a flat bottom bounded by potential "walls" which form the surface barrier. A free electron cannot absorb photons, since the laws of energy and momentum conservation cannot be simultaneously obeyed. Correspondingly, within the framework of Sommerfeld's model, absorption of light quanta by electrons is possible only in the surface layer owing to the existence of the potential drop (barrier). Somewhat more realistic models allow absorption to occur also in the metal bulk in view of the spatial changes of potential created by the lattice and electron field. According to the terminology suggested first by Tamm and Shubin (1931), these models divide the process of photoemission into bulk and surface effects. In particular, detailed model calculations of both photoexcitation mechanisms were carried out by Schaich and Ashcroft (1970, 1971) and Mahan (1970). The part of the "third body" (with respect to the electron and photon) in the bulk can also be played by impurities, dislocations, and lattice vibrations (phonons). Generally, the interaction ensuring the possibility of photoexcitation of the electron may be rather complex. For example, Endriz and Spicer (1971) have discussed a mechanism which assumes excitation by light of plasmons, i.e., collective vibrations of conduction electrons, with subsequent decomposition of plasmons and energy transfer to emitted electrons (a more detailed discussion is given in Section 10.4).

Possible mechanisms of scattering of light through excited electrons in metals will not be discussed further here. It should be mentioned only that owing to the presence of inelastic processes, the wave function of excited electrons is effectively attenuated outside the photoexcitation region. Therefore, regardless of the actual mechanism, the existence of a certain surface layer which plays the main role in processes connected with the photo-emission effect can be assumed. In particular, under conditions of photo-excitation in the bulk, the surface-layer thickness is determined by the smaller of the two quantities: the characteristic length of attenuation of the electron wave function caused by inelastic processes and the length of attenuation of the electromagnetic wave in the metal. If the former length is smaller, the surface layer is often called the "escape depth of electrons."

Early studies of photoemission from metals assumed the surface photo-effect to be dominant in the frequency range $\omega_0 < \omega < \omega_*$, and the bulk photoeffect, in the range $\omega > \omega_*$, corresponding to the penetration of the electromagnetic field into the metal. Subsequently, this simple treatment was held doubtful and the prevailing opinion regarded the bulk mechanism of photoexcitation to be dominant over the entire frequency range. However, the arguments advocating the latter view were in turn criticized. Present concepts consider that a sharp division of frequency range into regions corresponding to the surface and volume photoemission effect cannot be made in the general case. Physically, the reason lies in the close relation between surface and bulk processes, and particularly in the spatial de-localization of photoelectrons.

A more realistic, and physically valid procedure, is to specify not the coordinate but the energy ranges characterizing the escaping electrons. Proceeding from the final energy values of photoelectrons, two frequency ranges can be distinguished: near-threshold and extra-threshold frequencies (Brodskii and Gurevich, 1973). The main contribution to photoemission in the near-threshold region, which includes frequencies close to ω_0, is due only to those electrons which escape from the metal surface without ex-periencing other additional inelastic interactions. The energy losses suffered by photoexcited electrons in a single act of interaction are so large that, in the near-threshold frequency range, scattered electrons cannot escape from the metal for energetic reasons. The energy of the near-threshold frequencies is in the range of several electron volts. Quantitatively, the contribution of the surface and bulk mechanisms of photoexcitation in this region depends on the nature of the metal, its surface purity, the presence of bulk impurities and dislocations, etc.

The extra-threshold frequency region is characterized by the majority of emitted electrons having undergone interactions in the bulk metal, unrelated directly to photoexcitation. In this region, the photocurrent can contain a contribution from secondary electrons, which have gained energy in the process of interaction with primary, i.e., "original," photoelectrons. Simultaneously multiparticle effects, as well as band transitions and band-transition-determined properties of the dispersion of the dielectric constant of the metal, become more important. Thus, it is obvious that the laws of photoemission, within the extra-threshold frequency region, depend considerably on the actual properties of the given metal.

Conversely, in the vicinity of the photoelectric threshold, a frequency range exists in which a number of photoemission laws depend only on the general properties of metals and therefore have a generally universal application. Physically, the possibility of a unique threshold description lies in the fact that the kinetic energy of emitted electrons is lower than the energy parameters characterizing the internal structure of the emitting metal. It is obvious that the near-threshold region is of greatest importance for the purpose of investigating physicochemical processes at phase boundaries (but not for studying the electronic properties of the given metal). The extent of the electrode potential changes which can be applied in electrochemical systems also usually allows the behavior of photoelectrons to be controlled mainly in the near-threshold region.

1.2. The Threshold Approach to Photoemission

In accordance with the conclusions from the above discussion, investigation of electrochemical systems should be carried out mainly in the near-threshold frequency range (visible and near-ultraviolet part of the spectrum). In the paragraphs which follow, the principles of the threshold theory are discussed, which allow the photoelectron emission effect to be described in this frequency range.

Let the surface of a metal occupying the half-space $x < 0$ (Fig. 1.1) be irradiated by monochromatic light of frequency ω. If the quantum energy $\hbar\omega$ exceeds the value of the work function w of the electron escaping from the metal into the external medium, single photon emission becomes energetically possible [n-photon emission is significant if $(n - 1)\hbar\omega < w < n\hbar\omega$]. The photoelectric threshold is, by definition, given by $\hbar\omega_0 = w$.

The photoemission process is discussed here for stationary conditions, independent of the conditions at the beginning of experiment. The basic assumption concerning low final energy (in the sense discussed above) of emitted electrons is supplemented by the additional assumptions listed below.

1. The field of the external electromagnetic wave is sufficiently small that: (a) its strength can be neglected in comparison with that of interatomic fields; (b) its effect on the energy levels of emitted electrons outside the

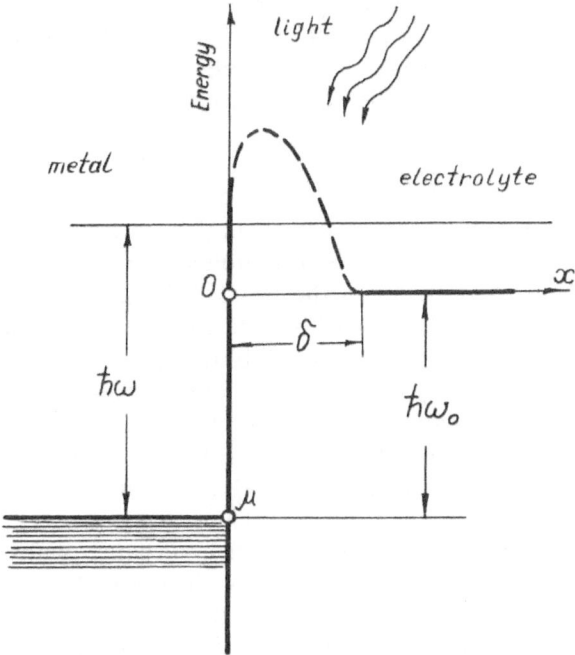

Fig. 1.1.   The energy diagram for a phase boundary.

emitter can also be neglected. Both conditions prevail for field strengths of the electromagnetic wave up to 10^6–10^7 V-cm^{-1} in order of magnitude.

2. The photocurrent is sufficiently small that the thermodynamic equilibrium within the solid remains virtually undisturbed. This is always true for ordinary experimental current densities.

3. The effect of the magnetic field of the incident wave on the behavior of electrons can be neglected. For the energy of quanta under consideration, when $\hbar\omega \ll m_0 c^2$ (c is the light velocity), this is also always possible (Landau and Lifshitz, 1959).

The general expression for the density of the photoemission current I directed perpendicularly to the infinite homogeneous emitter surface is then given by

$$I = e \int j_x(E_i, \mathbf{p}_\parallel, \omega) F(E_i, \mu) \rho(E_i, \mathbf{p}_\parallel)\, dE_i\, d\mathbf{p}_\parallel \qquad (1.1)$$

where the E_i and $\mathbf{p}_\parallel = \{p_y, p_z\}$ variables are the energy and parallel (to the surface) components of the momentum (quasi-momentum) of the initial electrons in the metal, respectively. (Here and later, "electrons" in the metal are to be understood as quasi-particles, the charge carriers e.)

The first factor within the integral in Eq. (1.1), $j_x(E_i, \mathbf{p}_{\parallel}, \omega)$, denotes the partial flux of emitted electrons, which upon multiplication by e gives the absolute value of the partial electric photocurrent corresponding to the initial values of E_i and $\mathbf{p}_{\parallel}$ for the original electrons. The second factor within the integral

$$F(E_i, \mu) = [e^{(E_i - \mu)/kT} + 1]^{-1} \tag{1.2}$$

describes the Fermi distribution of the initial electrons inside the metal. Here T is the absolute temperature, k is Boltzmann's constant, and μ is the chemical potential of electrons in the metal. Choosing the potential energy of the delocalized emitted electron outside the metal (beyond the effective range of surface forces, Fig. 1.1) as the arbitrary zero, we obtain $\mu = \hbar\omega_0 = -w$. The third factor, $\rho(E_i, \mathbf{p}_{\parallel})$, is the probability density function of initial states, which (as shown later) usually cannot be defined.

It follows from the definition of $F(E_i, \mu)$ and $\rho(E_i, \mathbf{p}_{\parallel})$ that the product $F(E_i, \mu)\rho(E_i, \mathbf{p}_{\parallel}) \, dE_i \, d\mathbf{p}_{\parallel}$ is the number of electrons in the metal for the range $dE_i \, d\mathbf{p}_{\parallel}$. Multiplying this quantity by $ej_x(E_i, \mathbf{p}_{\parallel}, \omega)$ and integrating over all allowed values of E_i and $\mathbf{p}_{\parallel}$ we obtain, in accordance with Eq. (1.1), the total photocurrent density. (For brevity, we shall not specify below whether the current from the total emitter surface or the current density is considered.)

The integration range in Eq. (1.1), i.e., the range of allowed E_i and $\mathbf{p}_{\parallel}$ values, is determined by the laws of energy and momentum conservation:

$$E_i + n\hbar\omega = \frac{1}{2m}(p^2 + \mathbf{p}_{\parallel}^2) \tag{1.3}$$

where n is the number of absorbed quanta of frequency ω, m is the effective mass corresponding to the motion of the emitted electron in the medium outside the metal (in vacuum $m = m_0$), and p is the xth momentum component for the emitted electron far from the boundary. It is assumed here that energy losses of photoexcited electrons are negligible within the metal. Equation (1.3) takes into account the fact that, owing to the translational symmetry in the boundary plane, values of p_y, p_z, the tangential momentum (quasi-momentum) components, are conserved when the electron crosses the boundary. From Eq. (1.3) we obtain $p = [2m(E_i + n\hbar\omega) - \mathbf{p}_{\parallel}^2]^{1/2}$. For the given values of n and ω, only those initial values of E_i and $\mathbf{p}_{\parallel}$ in Eq. (1.1) are allowed for which the radicand in the above expression is not negative and the momentum p is a real quantity (in the opposite case, p is purely imaginary and the corresponding current is equal to zero). Thus, the integration in Eq. (1.1) should be carried out over E_i and $\mathbf{p}_{\parallel}$ values which obey the condition

$$2m(E_i + n\hbar\omega) - \mathbf{p}_{\parallel}^2 > 0 \tag{1.4}$$

The central problem of the self-consistent quantum-mechanical theory is the calculation of the quantity $j_x[\psi_f]$ which, according to the general expression for the quantum-mechanical current operator, is given (Landau and Lifshitz, 1974) by

$$j_x[\psi_f] = \frac{\hbar i}{2m}\left(\psi_f \frac{\partial \psi_f^*}{\partial x} - \psi_f^* \frac{\partial \psi}{\partial x}\right) \tag{1.5}$$

where ψ_f is the time-independent wave function of the electron in its final state far from the emitter surface, ψ_f^* is the complex conjugate function to ψ_f, and $i = \sqrt{-1}$. Thus, j_x can be found only if the solution of the corresponding Schrödinger equation is available; moreover, the solution outside the metal depends, through conditions at the boundary, on the solution inside the metal.

The well-known theory of Fowler (1931) for the single-photon photoeffect does not explicitly involve the quantum-mechanical current; a semiclassical flux of electron gas incident from inside onto the metal surface is discussed. The theory of Fowler contains a number of additional phenomenological assumptions; essentially, however, it reduces (as will become obvious below) to a simple replacement in Eq. (1.1) of the quantity j_x by a constant. The proponents of several variants of the theory, developed mainly for the purpose of describing photoemission at relatively high frequencies, arrive at the value of ψ_f, and subsequently j_x, by solving model quantum-mechanical problems. The motion of electrons in the metal is described using the "box" model with various, and sometimes very complex, shapes of the "bottom" and "walls" (Adawi, 1964; Mahan, 1970; Schaich and Ashcroft, 1971).

Experimental data concerning photoemission into vacuum at light frequencies close to the threshold frequency ($|\omega - \omega_0)/\omega_0| \ll 1$) are in good agreement with Fowler's theory. The latter predicts that starting with energies of the order of a few kT above the threshold, $I \propto (\omega - \omega_0)^2$. At present, relations which follow from Fowler's theory are universally accepted for describing photoemission into vacuum in the threshold frequency range, although it became clear long ago that the theory is based on a series of insufficiently founded assumptions.

Attempts at using Fowler's formulas for describing photoemission into electrolyte solutions were unsuccessful (Barker *et al.*, 1966; Delahay and Srinivasan, 1966), and the ways in which modifications of existing calculation methods can be made remain totally obscure.

The approach to calculation of the quantity j_x, described below (Gurevich *et al.*, 1967; Brodskii and Gurevich, 1973), differs in principle from the methods mentioned above; it uses the calculation methods of quantum mechanics for threshold creation phenomena (Baz' *et al.*, 1971; Newton, 1967).

Fields affecting electron motion in the immediate vicinity of the surface are concentrated in the "transition" region of thickness δ (Fig. 1.1). The magnitude of δ is obviously close to interatomic distances in the metal. Consider first the region $x > \delta$. It is sufficiently far from the metal surface that the electron can be treated as if it were propagating in a field with a homogeneous effective potential $V(x)$ (which takes into account long-range fields). The unknown function ψ_f obeys the Schrödinger equation in the range $x > \delta$

$$\left[\frac{\hbar^2}{2m} \Delta + E_f - V(x) \right] \psi_f(x, y, z) = 0 \tag{1.6}$$

where $E_f \equiv E_i + \hbar\omega$ (unless otherwise stated, below we are concerned, for simplicity, with the case of single photoemission, i.e., when $n = 1$). The potential $V(x)$ becomes zero as $x \to \infty$. Taking into account that the quantity $\mathbf{p}_\parallel$ is conserved when the electron crosses the interface and choosing, for simplicity, the dependence of ψ_f on y and z given by

$$\psi_f = \exp\{i(\mathbf{p}_y y + \mathbf{p}_z z)/\hbar\}\psi(x)$$

the following basic expression is obtained, using Eq. (1.6):

$$\left[\frac{\partial^2}{\partial x^2} + \frac{p^2}{\hbar^2} - \frac{2m}{\hbar^2} V(x) \right] \psi(x) = 0 \tag{1.7}$$

According to the physical formulation of the problem, the desired solution should describe an electron spreading from the emitter surface. Owing to scattering processes in the external medium, e.g., in an electrolyte solution, the unknown wave function is, generally speaking, attenuated as $x \to \infty$. However, if the De Broglie wavelength of the emitted electron, $\lambda \equiv \hbar/p$, is smaller than the attenuation distance, the quantum-mechanical description of the act of photoemission can be made in the same way as in the absence of attenuation, e.g., in vacuum. Therefore, it will be assumed that for the region $x > \delta$, $\psi(x) = \mathscr{L}(p)f(x, p)$. Here $\mathscr{L}(p)$ is independent of x, and $f(x, p)$ is the solution of Eq. (1.7) which, as $x \to \infty$, describes the wave propagating from the metal surface and is normalized by the condition $j_x[f] = p/m$, where $j_x[f]$ is determined by Eq. (1.5). In particular, if the potential $V(x)$ tends exponentially (or faster) to zero as $x \to \infty$, then $f(x, p) = \exp(ipx/\hbar)$ obtains (with an accuracy comparable with the insignificant phase coefficient).

The solution of Eq. (1.7) obtained in this way is called the Jost solution.

For the chosen form of the function $f(x, p)$, we obtain upon substitution of ψ_t into Eq. (1.5) the following expression for j_x:

$$j_x[\psi_t] = \frac{p}{m} |\mathscr{L}(p)|^2 \tag{1.8}$$

It follows that the calculation of photoemission current reduces now to the determination of the squared modulus of the quantity $\mathscr{L}(p)$ which is a function of the final momentum p of the emitted electron and the light frequency ω, and a function of the potential $V(x)$.

Consider the possibility of a general determination of the functional relations described in the threshold approximation. If the transition layer thickness δ is sufficiently small, the coupling of the solution $\psi(x)$ of Eq. (1.7) at the boundary with the wave function at $x < \delta$ can be carried out using the value of $\psi(x)$ at $x = 0$ instead of at $x = \delta$. In order to assess this approximation qualitatively, the solution $\psi(x)$ in the vicinity of $x = \delta$ will be expanded in a Taylor series in the form

$$\psi(0) = \psi(\delta) - \frac{\partial \psi}{\partial x}\bigg|_{x=\delta} \delta + \frac{1}{2} \frac{\partial^2 \psi}{\partial x^2}\bigg|_{x=\xi} \delta^2 + \cdots$$

(point $x = \xi$ being contained within the range $[0, \delta]$).

The expansion of $|\psi(\delta)|^2$ by $|\psi(0)|^2$ is obviously valid if $|\psi(\delta) - \psi(0)|/|\psi(0)| < 1$.

It is clear from the above expansion and the relation between $\psi(x)$ and $d^2\psi/dx^2$ (Schrödinger equation) that the latter inequality always obtains if

$$\left|\frac{d \ln \psi}{dx}\right|_{x=\delta} \delta < 1, \qquad \frac{2m\delta^2}{\hbar^2} |V|_{\max} < 1 \tag{1.9}$$

where $|V|_{\max}$ is the maximum value of $|\Delta V|$ in the range $[0, \delta]$. It should be mentioned that the second condition in Eq. (1.9) can be more precisely defined and expressed as $2m|\Delta V|\delta^2/\hbar^2 \ll 1$, where $|\Delta V|$ is the largest deviation of the actual interaction in the range $[0, \delta]$ from that considered in the potential term in Eq. (1.7) and extrapolated to the same range.

In particular, if the total potential $V(x)$ drops exponentially (or faster) within a distance of the order of δ, we can equate $V(x)$ in Eq. (1.7) to zero. For $x > \delta$, the solution $\psi(x)$ turns out to be proportional to $\exp(ipx/\hbar)$, since the first condition in (1.9) can be expressed simply as

$$p\delta/\hbar \ll 1 \tag{1.10}$$

The latter condition has a clear physical interpretation: The thickness of the transition region, δ, should be small compared with the De Broglie wavelength of the emitted electron, $\lambda = \hbar/p$.

Assuming the conditions formulated above to be satisfied, consider now the region $x < 0$ inside the metal. The final energy of the escaping electron, E_f, appears in the corresponding equation of motion in the internal region, $x < 0$, only as a sum $(E_f + V_M)$, the interaction energy inside the metal, V_M, being much larger in its absolute value. Thus, over a sufficiently wide energy range ΔE_f, the quantity E_f varies much less than $[V_M]$. Therefore, if no separate energy levels exist in the metal (bulk or surface) within the variation of E_f considered, the solution ψ_f in the internal region should remain constant for small (as compared to V_M) changes of E_f. In other words, the quantity ψ_f is approximately independent of E_f in the threshold energy range ΔE_f and is equal to the value of $|\psi_f|$ corresponding to $E_f = 0$. The quantity $|\psi_f|$ is obviously independent of changes of E_f also at $x = 0$; therefore, $|\psi_f|_{x=0} = $ const, where "const" is a constant depending on the properties of the metal and is independent of the energy characteristics of the electron outside the metal. The latter equality should be used as a boundary condition for Eq. (1.7) at the metal surface. Energy values characteristic of interactions inside the metal (e.g., "depth of the potential well") are of the order of the kinetic energy of electrons at the Fermi surface E_F. (It should be mentioned that E_F is measured from the bottom of the conduction band and its order of magnitude is 10 eV.) It follows that the threshold treatment is valid if [together with Eq. (1.9)] the inequality $\Delta E_f / \Delta E_F \ll 1$ obtains. The order of magnitude of the final energy of electrons giving rise to the major part of the photoemission current at a light frequency ω does not exceed $\hbar(\omega - \omega_0)$. Correspondingly, the initial energies of these electrons are contained in the energy "layer" extending to the thickness $\hbar(\omega - \omega_0)$ near the Fermi surface of the metal. Therefore, the above inequality can be rewritten in the form

$$\frac{\hbar(\omega - \omega_0)}{E_F} \ll 1 \qquad (1.11)$$

Taking into account that ω_0 is the threshold frequency, the sense of the term "threshold approximation" can be understood from Eq. (1.11): The frequency ω should be sufficiently close to ω_0. Then, the initial energies of emitted electrons are close to that of the Fermi surface of the metal, and the probability of photoexcitation can be assumed approximately to be the same for all electrons and equal to the probability of photoexcitation at the Fermi surface. Differences in the behavior of electrons appear only outside the metal at the potential $V(x)$, which cannot be assumed to be much higher than the near-threshold energy range.

Assuming conditions (1.9) and (1.11) to be satisfied, and substituting $x = 0$ in the expression $\psi(x) = \mathscr{L}(p)f(x, p)$, we obtain $\mathscr{L}(p)f(x, p)|_{x=0} = $ const $= \Lambda$. Finally, using Eq. (1.8), the following expression for j_x results:

$$j_x = \frac{p}{m} |\Lambda|^2 / |f(p)|^2 \qquad (1.12)$$

where $f(p) = f(x, p)|_{x=0}$.

Equation (1.12), together with Eq. (1.7) which depends only on the potential $V(x)$ outside the metal, allows j_x (correct to within the constant factor) to be calculated, thus giving a solution to the problem.

The function $f(p)$ in Eq. (1.12) coincides, within the value of the significant phase factor, with the Jost function (Jost, 1947; Alfaro and Regge, 1965) well known in quantum-mechanical scattering theory. This often allows the results obtained by the theory of scattering to be directly used in the theory of photoemission.

It should be mentioned that the threshold approach discussed above leads to a solution correct to within a certain constant. One boundary condition is not sufficient to calculate completely the current j_x, and the magnitude of $|\Lambda|^2$ remains unknown. However, for many cases it is sufficient that $|\Lambda|^2$ be independent of p for the dependence of j_x and, consequently, of the total current I on the form of the force fields outside the metal to be obtained.

1.3. Calculation of the Photoemission Current

The total photoemission current density can be found by substitution of the expression for $j_x(E_1, \mathbf{p}_\parallel, \omega)$ into Eq. (1.1) and integration over the initial states of electrons in metal, determined from Eq. (1.4). The upper limit of integration (over energy E_1) equals infinity, the lower limit, $\hbar\omega$.

The threshold condition (1.11) simplifies the calculation considerably. In the first place, at ordinary temperatures, the number of filled initial states with energy higher than the Fermi level (i.e., $E_1 > \mu$) decreases sharply, the cut-off factor being the function $F(E_1, \mu)$. The lower limit of the integral in Eq. (1.1) is due to the existence of the energy threshold, and is also close to μ [cf. Eq. (1.11)]. Therefore the main contribution to I is due to electrons with initial energy E_1 close to μ. Similarly, the allowed values of $\mathbf{p}_\parallel$ are contained in a narrow region in the vicinity of $p_\parallel = 0$ [cf. Eq. (1.4)].

Thus, Eq. (1.1) can be integrated assuming the density of electron states, $\rho(E_1, \mathbf{p}_\parallel)$, to be constant and equal to $\rho(E_1, \mathbf{p}_\parallel) \simeq \rho_0$, where $\rho_0 \equiv \rho(E_1, \mathbf{p}_\parallel)|_{E=\mu, \mathbf{p}_\parallel = 0}$. (It is assumed that the Fermi surface of the metal contains the point $\mathbf{p}_\parallel = 0$ and is sufficiently smooth in the vicinity of this point; cf. also Section 4.3.) Furthermore, the current j_x depends (in the threshold approximation) on the combination rather than on the separate values of

$E_\perp$ and $\mathbf{p}_\parallel$ which appear in the expression for p. In turn, p depends only on $\mathbf{p}_\parallel{}^2 = |\mathbf{p}_\parallel|^2$. Therefore, Eq. (1.1) can be immediately integrated once to obtain $d\mathbf{p}_\parallel = 2\pi|\mathbf{p}_\parallel|d|\mathbf{p}_\parallel|$. Introducing, in place of $E_\perp$ and $|\mathbf{p}_\parallel|$, the new variables $E_\perp$ and $E \equiv p^2/2m$, we can integrate over $E_\perp$ for the general case. In the final result, the absolute value of I (Brodskii and Gurevich, 1973) is given by

$$I = 2\pi e \rho_0 mkT \int_0^\infty j_x(\sqrt{2mE}) \ln \left[1 + e^{(\mu + \hbar w - E)/kT} \right] dE \qquad (1.13)$$

It should be mentioned that Eq. (1.13) was derived without any assumptions concerning the electron dispersion law for metals.

At ordinary temperatures $kT \ll E_F$ is always valid for metals. Therefore, it can be assumed that $T = 0$ at frequencies such that $kT \ll \hbar(\omega - \omega_0) < E_F$. Using the relation

$$\lim_{T \to 0} \{kT \ln(1 + e^{y/kT})\} = y\theta(y)$$

where $\theta(y)$ is a step function

$$\theta(y) = \begin{cases} 1 & \text{for} \quad y > 0 \\ 0 & \text{for} \quad y < 0 \end{cases}$$

we obtain from Eq. (1.13)

$$I = 2\pi e \rho_0 m \int_0^{\hbar(\omega - \omega_0)} j_x(\sqrt{2mE})(\hbar\omega - \hbar\omega_0 - E) \, dE \qquad (1.14)$$

Relations (1.13) and (1.14), together with (1.12) in the general case, allow the problem of calculating the photoemission current in the near-threshold range of frequencies to be solved. Assessments of the range of their validity show that for δ of the order of 1–2 Å, the range of threshold energies amounts to 1–1.5 eV. This corresponds to the usual range of electrode potential variations (in volts) in electrochemical measurements.

Using the relations derived above, it is useful to consider now the photoemission of electrons into a solid or liquid dielectric. Generally speaking, in this case, for the range $x > \delta$, the electron moves in the field of image forces due to the surface charge on the metal introduced by the emitted electron itself. The potential due to these forces is equal to $V(x) = -e^2/4\varepsilon x$, where ε is the dielectric constant of the external medium; the value of the potential for $\varepsilon = 1$ corresponds to emission into vacuum. In the present case, Eq. (1.7) becomes

$$\left(\frac{\partial^2}{\partial x^2} + \frac{p^2}{\hbar^2} + \frac{me^2}{2\hbar^2 \varepsilon x} \right) \psi(x) = 0 \qquad (1.15)$$

Equation (1.15) coincides with the well-known equation describing the motion of a charge with zero orbital momentum in a Coulombic field. Solutions of the latter equation, $\mathscr{G}_0$ and $\mathscr{F}_0$, are also well known (cf., e.g., Goldberger and Watson, 1964, Section 6.4). For the coulomb potential, the solution describing the emitted wave is of the form $f(x, p) = \mathscr{G}_0 + i\mathscr{F}_0$. After substitution in Eq. (1.12), we obtain

$$j_x = \frac{p_e}{m}|\Lambda|^2 \left|1 - \exp\left(-\frac{p_e}{p}\right)\right|^{-1} = \begin{cases} \dfrac{p_e}{m}|\Lambda|^2 & \text{for} \quad p \ll p_e \\[2mm] \dfrac{p}{m}|\Lambda|^2 & \text{for} \quad p \gg p_e \end{cases} \tag{1.16}$$

where $p_e = \pi e^2 m/2\varepsilon\hbar$. The corresponding energy, $E_e \equiv p_e^2/2m$, is given in atomic units of energy and is equal to $E_e = (\pi^2/8\varepsilon^2)(m/m_0)(m_0 e/\hbar^2)^4$, or numerically

$$E_e = \frac{33.5}{\varepsilon^2}\frac{m}{m_0} \text{ eV} \tag{1.17}$$

The quantity E_e thus determined is characteristic of the external medium.

The general expression for the photoemission current density, I, which depends on two dimensionless parameters $\beta \equiv \hbar(\omega - \omega_0)/kT$ and $j \equiv \hbar(\omega - \omega_0)/E_e$, can be obtained by substitution of Eq. (1.16) into Eq. (1.13). Introducing a dimensionless variable $y \equiv E/kT$ we obtain

$$I = A_0 T^2 \zeta |\Lambda|^2 \sqrt{\frac{E_e}{E_F}} \int_0^\infty (1 - e^{-\sqrt{\beta/\gamma y}})^{-1} \ln(1 + e^{\beta - y}) \, dy \tag{1.18}$$

where $A_0 \equiv 4\pi k^2 e m_0/(2\pi\hbar)^3$ is the Sommerfeld constant ($A_0 = 120.4$ A-cm^{-2}-deg^{-2}), E_F is the Fermi energy of the metal [which appears in Eq. (1.18) as a result of transformation of the expression for ρ_0 based on the free electron model], and ζ is a dimensionless function which characterizes the metal and describes the deviation of the actual statistical behavior of electrons in the metal from that of an ideal Fermi gas (for the latter, $\zeta = 1$).

We now consider photoemission into a dielectric with a relatively large value of the parameter $(m/m_0)\varepsilon^{-2}$ so that $E_e > E_F$. In particular, this case corresponds to emission into a vacuum.

In the threshold frequency range we have $\gamma \ll 1$, and the quantity $e^{-\sqrt{\beta/\gamma y}}$ in Eq. (1.18) can be neglected since it is much smaller than unity. Introducing a new variable $u = \beta - y$ and defining $\zeta|\Lambda|^2\sqrt{E_e/E_F} \equiv \alpha_F$, we obtain the expression for the photoemission current,

$$I = A_0 T^2 \alpha_F \int_{-\infty}^\beta \ln(1 + e^u) \, dU \tag{1.19}$$

which coincides with the final formula of the Fowler theory (Fowler, 1931; Dobretsov and Gomoyunova, 1966). Thus, the numerous model-based assumptions commonly used in the derivation of Eq. (1.19) are equivalent to a single assumption: the phenomenological postulation of the condition j_x = const. In accordance with Eq. (1.16), it turns out that in the case of photoemission into vacuum the relation $j_x = (p_e/m)|\Lambda|^2$ = const is indeed valid, owing to the significance of image forces (with very general assumptions concerning the metal structure).

It follows from the method of derivation of Eqs. (1.18) and (1.19) that they describe not only single but also n-photon (i.e., $n\omega \simeq \omega_0$) photoemission of electrons near the corresponding threshold. In particular, for $\beta \gg 1$, we obtain from Eq. (1.19)

$$I = \frac{A_0 \alpha_F}{2k^2} \hbar^2 (n\omega - \omega_0)^2 \qquad \text{for} \qquad n\omega > \omega_0 \qquad (1.20)$$

i.e., the photocurrent increases with distance from the threshold according to a square law. For $n = 1$, Eq. (1.20) can be transformed into $I \propto (\omega - \omega_0)^2$, a relation already mentioned earlier.

If $E_e > E_F$, the parameter γ in Eq. (1.18) can be larger or smaller than unity, already within the threshold-energy range. This case may correspond experimentally to emission into dielectrics. Assuming $\beta \gg 1$ and introducing a new variable $u = y/\beta$, we obtain from Eq. (1.18)

$$I = A_0 \frac{\alpha_F}{2k^2} \hbar^2 (\omega - \omega_0)^2 G(\gamma) \qquad \text{for} \qquad \omega > \omega_0 \qquad (1.21)$$

where the dimensionless function $G(\gamma)$ is of the form

$$G(\gamma) = \int_0^1 \frac{2(1-u)\,du}{1 - \exp\{-(\gamma u)^{-1/2}\}} = \begin{cases} 1 + 8\gamma \exp(-\gamma^{-1/2}) + \cdots & \text{for} \quad \gamma \ll 1 \\ \frac{8}{15}\gamma^{1/2} + \frac{1}{2} + \frac{2}{9}\gamma^{-1/2} + \cdots & \text{for} \quad \gamma \gg 1 \end{cases} \qquad (1.22)$$

Values of $G(\gamma)$ in the intermediate range ($\gamma \simeq 1$) are tabulated in Appendix 5, which also contains the plot $g(\gamma) \equiv \log G(\gamma)$.

It can be seen that for $\gamma \ll 1$, Eq. (1.22) reduces to Eq. (1.20); however, for $\gamma \gtrsim 1$, the spectral characteristics of the photocurrent deviate from the square-law dependence given in Eq. (1.20).

In conclusion, some general remarks should be made. In any case, the equations for the photoemission current (1.20) and (1.21) are valid for "normal" metals in the threshold range of the final energies of emitted electrons for both surface and bulk excitation provided the transition is an interband one. At the same time, the developed theory, naturally, cannot describe all the conceivable features of the photoelectron emission even in the near-threshold range. Thus, there may be significant deviations from Eqs. (1.20) and (1.21) due to the presence of the resonance energy levels (in part this problem was considered in Section 1.5). Moreover, some different relationships may be valid if bulk or surface plasmons play an important role in the generation of photoelectrons (see Section 10.4). In some cases narrow d bands may make an additional contribution to the photocurrent.

1.4. The 5/2 Power Law

The case of photoemission into an electrolytic solution is both the simplest and the most important one. The total potential drop in the system occurs virtually over the compact part of the double layer of thickness d (see, e.g., Fig. 0.2). The value of δ can be chosen so that $d < \delta$, and the entire double layer then lies within the range $0 < x < \delta$. Outside this region, $V(x) = 0$, since, as opposed to the case of vacuum emission, image forces are effectively screened in this case.

The quantum-mechanical description of screening effects requires solution of a separate, quite complex problem of multiple bodies which lies beyond the scope of a descriptive, single-particle treatment of electron motion in the external medium. Therefore, we shall restrict the discussion to qualitative remarks.

The theory is based on the time-independent Schrödinger equation (1.6), the electron being described by a monochromatic wave function (fixed E_f value). Formally, this corresponds to a strictly constant (in time) electron density distribution outside the emitting electrode. As opposed to the classical "escape" of isolated electrons, the quantized character of photoemission leads in this approximation to a stationary probability flux in the direction $x \to \infty$. Therefore, after the transition period corresponding to the beginning of the experiment, photoemission, in this approximation, should not be accompanied by spatial variations of the charge-density distribution. Correspondingly, ions present in the electrolyte should redistribute so as to screen the emitted charge.

Actually, the electron wave function is not strictly monochromatic and has a corresponding wave packet. Therefore, the screening time is not infinite (as in the case of a monochromatic wave), but less than the time τ_w the wave packet takes to cross the boundary. However, τ_w is still much longer than the time required for the classical point charge to cross the double layer. Moreover, by analogy to the dynamic charge screening in metals, collective motion of the type associated with plasma vibrations should appear. These collective motions (and not, for example, classical diffusion of ions) result in the effective elimination of long-range forces.†

Introducing $V(x) = 0$ for $x > \delta$ into Eq. (1.7), we obtain

$$\left(\frac{\partial^2}{\partial x^2} + \frac{p^2}{\hbar^2}\right) f(x, p) = 0 \qquad \text{for} \qquad x > \delta \qquad (1.23)$$

The solution of Eq. (1.23) is in the form $f(x, p) = \exp\{ipx/\hbar\}$, from which, according to Eq. (1.12),

$$j_x = \frac{p}{m} |\Lambda|^2 \qquad (1.24)$$

The total photoemission current I can be calculated by substitution of Eq. (1.24) into the general formula (1.13) or by considering the limiting case, $\gamma \to \infty$, in Eq. (1.18).

Integrating once by parts we obtain (Gurevich *et al.*, 1967)

$$I = \tfrac{2}{3} A_0 \zeta |\Lambda|^2 T^2 \sqrt{\frac{kT}{E_F}} \int_0^\infty \frac{u^{3/2}\, du}{\exp(u - \beta) + 1} \qquad (1.25)$$

where $\beta \equiv \hbar(\omega - \omega_0)/kT$; A_0 and ζ are constants independent of T and ω. A large-scale plot of the function

$$B(\beta) \equiv \int_0^\infty \frac{u^{3/2}\, du}{\exp(u - \beta) + 1} \qquad (1.26)$$

is given in Appendix 2 for the range -3.5 to 3.5. Outside this range, the asymptotic expression

$$B(\beta) = \begin{cases} 1.33 e^\beta & \text{for} \quad |\beta| \gg 1, \beta < 0 \\ 2(\beta/5)^{5/2}(1 + 5\pi/8\beta^2) & \text{for} \quad \beta \gg 1 \end{cases} \qquad (1.27)$$

applies with less than 1% error. For $T \to 0$ and $|\beta| \to \infty$, Eqs. (1.25)–(1.27) yield

$$I = \begin{cases} 0 & \text{for} \quad \omega < \omega_0 \quad (1.28a) \\ \tfrac{4}{15} A_0 \zeta |\Lambda|^2 k^{-2} E_F^{-1/2}(\hbar\omega - \hbar\omega_0) & \text{for} \quad \omega > \omega_0 \quad (1.28b) \end{cases}$$

† The above discussion corresponds to the following conditions: $\tau_w \simeq l_w/v$, where $v = 10^7$ to 10^8 cm-sec^{-1} is the velocity of the escaping electron and $l_w \simeq 10^{-6}$ cm is the characteristic dimension of the packet, whence $\tau_w \simeq 10^{-13}$ to 10^{-14} sec. The time characteristic of collective ion motions is of the order of $\tau_i = (4\pi e^2 N/M)^{-1/2}$, where N is the number of ions in unit volume and M their mass. For $N \simeq 10^{19}$ to 10^{20} cm^{-3} and $M \simeq 10^3$ to 10^4 m_0, τ_i is of the order of τ_e. The corresponding plasma frequency is given by $\omega_i = [4\pi e^2 N/M]^{1/2}$, where N is the number of charged particles per unit volume, M is their mass; for $N \simeq 10^{19}$ to 10^{20} cm^{-3} and $M = 10^3$ to $10^4 m_0$, ω_i is of the order of τ_w^1. The frequency ω_T connected with the energy of thermal vibrations, $\hbar\omega_T = kT$, is of the same order of magnitude.

It follows from Eq. (1.28) that ω_0 is indeed, in accordance with the strict definition, the photoelectric threshold. This can be demonstrated in a similar way for Eq. (1.19). In reality, when $T > 0$ the photocurrent can also be observed for $\omega < \omega_0$ [cf. Eq. (1.25)]. The current originates from those electrons inside the metal which, when $T > 0$, are thermally excited to energy levels higher than that of the Fermi surface. The temperature effect (the so-called "thermal tail") quickly decreases with increasing frequency ω, and when $\hbar(\omega - \omega_0) \gg kT$ the photoemission current is given by Eq. (1.28) derived for $T = 0$. Therefore, the threshold frequency ω_0 can be reliably determined by extrapolation to $I = 0$ in experiments carried out at $T = 0$.

It can be seen by comparison of Eqs. (1.18) and (1.25) that laws for photoelectron emission into vacuum (or into dielectrics with sufficiently low permittivity ε) and into concentrated electrolytic solutions differ considerably. In particular, for the experimentally important range $\beta \gg 1$, the first case obeys Eq. (1.20) while the second obeys Eq. (1.28).

Experimental results for photoemission into vacuum are often treated using the function

$$f(\beta) \equiv \log\left[\int_{-\infty}^{\beta} \ln(1 + e^u)\, du\right] \tag{1.29}$$

On the other hand, photoemission into electrolytes is often conveniently treated using the universal function

$$b(\beta) \equiv \log\left[\int_{0}^{\infty} u^{3/2}(1 + e^{u-\beta})^{-1}\, du\right] \tag{1.30}$$

which is plotted in Appendix 2, together with the function $f(\beta)$ [cf. Eq. (1.29)].

The dependence of I on the electrode potential φ in concentrated electrolytic solutions can be found using the expression for the work function in solution [Eq. (0.1)], rewritten here in the form

$$\hbar\omega_0(\varphi) = \hbar\omega_0(0) + e\varphi \tag{1.31}$$

where $\omega_0(0)$ is the photoelectric threshold at $\varphi = 0$. The choice of zero potential is arbitrary (e.g., it can be the potential of a reference electrode or the potential of zero charge, etc.). The work function, however, depends on this choice. It follows from Eq. (1.31) that the photoelectric threshold varies with the applied potential φ.

Thus, Eqs. (1.25) and (1.31) fully describe the voltametric characteristics of the metal electrode–concentrated electrolyte solution system under conditions of photoelectron emission. The photoemission current I depends on the sum $\hbar\omega - \hbar\omega_0(0) - e\varphi$ and thus varies appreciably with the electrode potential φ even for a fixed value of the light frequency ω.

Substitution of the latter sum into the limiting expression (1.28b) leads in the extra-threshold frequency range to a relationship called (Gurevich *et al.*, 1967) the "5/2 power law":

$$I = A(\hbar\omega - \hbar\omega_0(0) - e\varphi)^{5/2} \tag{1.32}$$

where $A \equiv \frac{4}{15}A_0\zeta|\Lambda|^2 k^{-2} E_F$ is the expression preceding the brackets in Eq. (1.28b).

It can be seen from the derivation of Eq. (1.32) that the quantity $E_m \equiv \hbar\omega - \hbar\omega_0(0) - e\varphi$ equals (neglecting thermal effects) the maximum possible kinetic energy of emitted electrons for the given values of φ and ω.

The 5/2 power law, (1.32), describes the voltammetric characteristics of the system in the near-threshold energy range for single- and multiphoton effects (in the latter case, ω is replaced by $n\omega$). It is the basic relation for the analysis of photoemission phenomena in electrochemical systems.

1.5. The Effects of the Double Layer on the Photoemission

The 5/2 power law (1.32) was derived assuming the transition region $[0, \delta]$ to be sufficiently "narrow" that conditions (1.9) are fulfilled. This assumption may, however, become invalid in the case of extended thickness of the double layer (e.g., in dilute solutions or in the presence of "big" molecules adsorbed at the electrode) as well as for higher energies of the escaping electron (i.e., for smaller De Broglie wavelengths, $\lambda = \hbar/p$). Under these conditions, deviations from the 5/2 power law, and even qualitative changes of the voltammetric curve, can be expected. In the physical sense, these deviations are due to the following: If the quantity λ is commensurate with the characteristic distance over which the potential drop occurs in the vicinity of the electrode, the electron "feels," as it were, the details of the potential change. The photoemission behavior is then similar to that of an electron source in an unusual type of "electron microscope" which allows the electrons to be used for "examination" of the structure of the interface within atomic distances (cf. field emission microscopy).

The quantitative relations describing the dependence of the photoemission current on potential will be derived below for some special cases.†

The Compact Part of the Double Layer (Helmholtz Layer)

Deviations from the 5/2 power law due to increasing values of $E_m \equiv \hbar\omega - \hbar\omega_0(0) - e\varphi$ will be considered first. Model considerations are spurious

---

† The text which immediately follows can be omitted by readers not interested in details of the calculation. The necessary final equations are repeated in the discussion of the corresponding experimental data.

in this case; it is enough to examine the general properties of the Jost function. When the distance over which the potential $V(x)$ drops can be included within the region $[0, \delta]$, so that $V(x) = 0$ for $x > \delta$, the function $f(p) = 1$ or, more generally, $|f(p)|^2 = \text{const}$. The latter equality can be shown to be the first term of the expansion of the function $|f(p)|^2$ (Alfaro and Regge, 1965):

$$|f(p)|^2 = \mathfrak{A} + \mathfrak{B}p^2 + \cdots \tag{1.33}$$

where $\mathfrak{A}$ and $\mathfrak{B}$ are p-independent constants. It will be assumed that the quantum energy $\hbar\omega$ is sufficiently high and the "distance" from the threshold sufficiently large that the next term in the expansion (1.33) becomes significant. Then assuming $\mathfrak{A} \gg \mathfrak{B}p^2$, we obtain

$$|f(p)|^2 = \frac{1}{\mathfrak{A} + \mathfrak{B}p^2} \simeq \frac{1}{\mathfrak{A}}\left(1 - \frac{\mathfrak{B}}{\mathfrak{A}}p^2\right) \tag{1.34}$$

and after substitution in Eq. (1.14),

$$I = AE_m^{5/2}\left[\frac{1}{\mathfrak{A}} - \frac{6m\mathfrak{B}}{7\mathfrak{A}^2}E_m\right] \tag{1.35}$$

Thus, the deviation from the 5/2 power law increases linearly with increasing light frequency ω regardless of the dependence of potential changes on the detailed structure of the interface.

Calculation of $\mathfrak{A}$ and $\mathfrak{B}$ can be carried out, for example, within the framework of the usual condenser model of the double layer of thickness d (Frumkin *et al.*, 1952; Delahay, 1965). The potential $V(x)$ is given in this case by

$$V(x) = \begin{cases} -e\varphi(1 - x/d) & \text{for} \quad 0 \leqslant x \leqslant d \\ 0 & \text{for} \quad x \geqslant d \end{cases} \tag{1.36}$$

(The potential of zero charge is chosen here as the arbitrary zero.) Using Eq. (1.36), the Schrödinger equation (1.7), and introducing a new variable

$$\zeta = \left(x - \frac{E + e\varphi}{e\varphi}d\right)(-2me\varphi/\hbar^2d)^{1/2}$$

the unknown solution $f(x, p)$ in the region $0 \leqslant x \leqslant d$ is given in the form

$$f'' + \zeta f = 0 \tag{1.37}$$

The general solution of Eq. (1.37) can be expressed by a linear combination of Airy functions Ai and Bi of the argument $-\zeta$ (Abramowitz and Stegun, 1965) so that for $f(x, p)$ we obtain

$$f(x, p) = \begin{cases} a\,\text{Ai}(-\zeta) + b\,\text{Bi}(-\zeta) & \text{for} \quad x \leqslant d \\ \exp(ipx/\hbar) & \text{for} \quad x \geqslant d \end{cases} \tag{1.38}$$

where a and b are integration constants determined from the continuity condition of functions $f(x)$ and df/dx at $x = d$. Calculation of integration constants and of $f(p) = f(x, p)_{x=0}$ shows that $f(p)$ depends (apart from p) on a single dimensionless parameter

$$\eta \equiv -2me\varphi d^2/\hbar^2$$

which obviously fully describes the effect of potential [Eq. (1.35)] on the escaping electron. In the limiting case when $pd/\hbar \ll 1$ and $|\eta| \ll 1$, the expressions for $f(p)$ are considerably simplified owing to the asymptotic forms of the Ai and Bi functions; then the general expression (1.34) takes the form (Sheberstov, 1970)

$$I = AE_m^{5/2}\left[1 - \tfrac{1}{3}\eta + \tfrac{2}{35}\eta \frac{md^2}{\hbar^2} E_m + \cdots\right] \qquad (1.39)$$

Equation (1.39) is a more precise model definition of Eq. (1.34); it can be also used for a more correct assessment of the range of validity of Eq. (1.32). It should be mentioned here that the quantity η becomes of the order of unity for $d \simeq 1$ to 1.5 Å and $|\varphi| \simeq 1.5$ to 2 V.

The Diffuse Part of the Double Layer (Gouy Layer)

We shall now discuss photoemission in relatively dilute electrolytic solutions and determine the dependence of the photoemission current on the ψ' potential.

The model of the double layer commonly used (cf. Section 0.3), consisting of compact and diffuse parts (Fig. 1.2), will be considered.

The potential at the inner limit of the latter part is called the ψ' potential. The charge in the double layer is averaged (smeared out) over the electrode surface both in the presence and absence of specifically adsorbed ions, allowing the model to be treated in one-dimensional terms. It should be mentioned that the applicability of a one-dimensional potential distribution (as opposed to a micropotential depending on three spatial coordinates) has been a much discussed and not yet fully resolved problem in electrochemistry. The case of photoelectron emission is simpler. Owing to its wave nature, the escaping electron is delocalized in the plane of the electrode surface, being subject to a potential averaged over this plane, thus validating the use of a one-dimensional model.

At relatively low ψ' values ($|e\psi'/kT| \ll 1$), the potential varies with distance in the $x > d$ region (d is the compact layer thickness) as follows:

$$V(x) = -e\psi' \exp\{-\kappa(x - d)\}$$

where

$$\kappa^{-1} \equiv \left(kT\varepsilon_0/4\pi e^2 \sum_i z_i^2 n_i\right)^{1/2} \qquad (1.40)$$

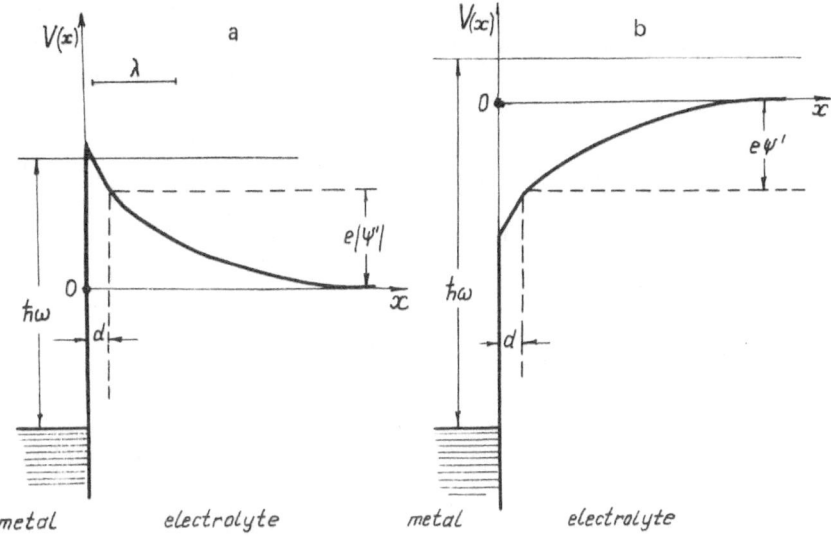

Fig. 1.2.   Potential drop in the double layer for dilute solutions. (a) $\psi' < 0$; (b) $\psi' > 0$.

is the thickness of the diffuse double layer, z_i and n_i are the charge and number of ions per 1 cm³ of the ith kind, respectively, and ε_0 is the static permittivity of the solution. If $|e\psi'/kT| \gg 1$, the potential distribution $V(x)$ requires a more complex analytical description (Delahay, 1965). In the case considered, however, Eq. (1.40) can also be used for an approximate description of the potential distribution for $|e\psi'/kT| \gg 1$. In fact, the field of the diffuse layer affects the escaping electrons primarily at the distance $x = d$, where the change of $V(x)$ is sharpest. The value of $V(x) = -e\psi'$ obtained from Eq. (1.40) coincides with that obtained from the exact expression. The quantity κ^{-1} [cf. Eq. (1.40)] can be used as an adjustable parameter to choose the value $\kappa = \kappa_*$ in such a way that the derivatives at $x = d$ coincide in both Eq. (1.40) and the exact expressions. For this purpose

$$\kappa_* = \kappa(2kT/e\psi') \sinh(e\psi'/2kT)$$

Expression (1.40) describes, to a good approximation, the effect on the escaping electron of the potential distribution in the diffuse layer for all ψ' values and it becomes an exact one for low ψ' values. Below, Eq. (1.40) is used to describe $V(x)$ for all ψ' values, assuming the effective value of κ to be given by $\kappa_* = \kappa(2kT/e\psi') \sinh(e\psi'/2kT)$.

The Schrödinger equation (1.7) for the given potential $V(x)$ can be rewritten in the form

$$\frac{d^2f}{dy^2} + \frac{1}{y}\frac{df}{dy} + \left[1 + \left(\frac{2p\kappa^{-1}}{\hbar y}\right)^2\right]f = 0 \qquad (1.41)$$

after introduction of a new variable $y = 2(\kappa^{-1}/\hbar)[-2mV(x)]^{1/2}$. Equation (1.41) has two independent solutions in the form of Bessel functions $\mathscr{J}_{i\nu}(y)$ and $\mathscr{J}_{-i\nu}(y)$, where $\nu = 2p\kappa^{-1}/\hbar$. Using the expansion (Abramowitz and Stegun, 1965)

$$\mathscr{J}_q = \sum_{m=0}^{\infty} (-1)^m \left(\frac{y}{2}\right)^{q+2m} \frac{1}{m!\,\Gamma(m+q+1)} \tag{1.42}$$

where $\Gamma(z)$ is the Euler gamma function, the asymptotic form of the function $\mathscr{J}_{\pm i\nu}(y)$ for $x \to \infty$ can be easily found for $y \to 0$:

$$\mathscr{J}_{\pm i\nu} = \left(\frac{2me\psi'}{\hbar^2\kappa^2}\right)^{\pm\nu/2} \frac{e^{\pm ip(x-d)/\hbar}}{\Gamma(1 \pm i\nu)} \qquad \text{as} \qquad x \to \infty$$

from which

$$f(x,p) = e^{-ipd/\hbar}\left(\frac{2me\psi'}{\hbar^2\kappa^2}\right)^{i\nu/2}\Gamma(1-i\nu)\mathscr{J}_{-i\nu}\left(\frac{2}{\hbar\kappa}(2me\psi')^{1/2}e^{-\kappa(z-d)/2}\right) \tag{1.43}$$

Let $\psi' < 0$. Consider first the case of a relatively dilute solution for which the De Broglie wavelength of the escaping electron is less than the diffuse-layer thickness, κ^{-1}; in this case, obviously, $\nu \gg 1$. Using a suitable asymptotic expression for the Bessel function with a purely imaginary index, for the electron energy range $E \equiv p^2/2m > |e\psi'|$, we have

$$f(x,p)|_{x=d} = \exp\left\{i\frac{d}{\hbar}[2m(E-|e\psi'|)]^{1/2}\right\}$$

$$\frac{df}{dx}\bigg|_{x=d} = \frac{i}{\hbar}[2m(E-|e\psi'|)]^{1/2}\exp\left\{i\frac{d}{\hbar}[2m(E-|e\psi'|)]^{1/2}\right\} \tag{1.44}$$

The same asymptotic expressions can be used to demonstrate that for $\nu \gg 1$ the contribution to the photocurrent from electrons with energies within $0 < E < |e\psi'|$ becomes exponentially smaller. For $\nu \ll 1$ and $E < |e\psi'|$, electrons freely penetrate the relatively narrow barrier of the diffuse layer. Therefore, the effect of the ψ' potential on photoemission is negligibly small in this case. It is even smaller for $E > |e\psi'|$. In other words, the effect of the diffuse layer on photoemission can be neglected in the range $\nu \ll 1$. The inequality

$$(2m|e\psi'|/\hbar\kappa^{-1})^{1/2} \ll 1$$

is sufficient to ensure these conditions.

It can easily be shown that the intermediate range (between the two ranges discussed $\nu \gg 1$ and $\nu \ll 1$), $\nu \simeq 1$, is in fact a very narrow one, so that conditions "much larger (smaller)" can be replaced by "larger (smaller)."

The relations (1.44) obtained constitute, in fact, the boundary conditions for the Schrödinger equation in the range $0 < x < d$ and allow the subsequent determination of $f(p) = f(x, p)|_{x=0}$. They describe fully the behavior of the photocurrent in the $x > d$ range and, consequently, the effect of the ψ' potential. Relations (1.44) correspond to those boundary conditions which would obtain for $V(x) = \text{const} = -e\psi' > 0$ when $x > d$.

In fact, by replacing $V(x)$ in Eq. (1.7) by $|e\psi'| = \text{const}$, we obtain for the $x > d$ range

$$f(x, p) = \exp\left\{\frac{ix}{\hbar} [(2mE - |e\psi'|)]^{1/2}\right\}$$

It can easily be seen that the values of $f(x, p)_{x=d}$ and $(df/dx)_{x=d}$ coincide with those given by Eq. (1.44).

Thus, the effect of the ψ' potential is insignificant for $\psi' < 0$ and $\nu < 1$, and for $\psi' < 0$ and $\nu > 1$ it reduces to the effective change of the reference energy level by $|e\psi'| > 0$. The latter result has a sufficiently obvious physical meaning. Condition $\nu > 1$ means that in the range $x > d$, significant changes of $V(x)$ occur only at distances much larger than the De Broglie wavelength of the emitted electron. Therefore, the diffuse layer plays the part of the "flat bottom box" in this case in a similar way that the uncharged electrolyte bulk does in concentrated solutions. In other words, the change of the electrode potential by φ lowers the bottom of the energy well for electrons emitted in the dilute solution by a quantity $|e(\varphi - \psi')|$, and not by $|e\varphi|$ as is the case for concentrated solutions. It follows that all the results obtained above apply to appropriately dilute solutions if the potential φ is replaced by $\varphi - \psi'$. In particular, Eq. (1.32) is replaced by (Sheberstov *et al.*, 1970)

$$I = A[\hbar\omega - \hbar\omega_0(0) - e(\varphi - \psi')]^{5/2} \tag{1.45}$$

which will be referred to subsequently as "the modified 5/2 power law."

Consider now the case $\psi' > 0$ (Fig. 1.2b). It is important because the photocurrent can become abnormally large under these conditions. The sharp increase of the photocurrent due to a certain configuration of the potential well will subsequently be called surface emission resonance.

If in expansion (1.33) the quantity $\mathfrak{A}$ equals zero, the photocurrent should, according to Eq. (1.12), increase sharply as $p \to 0$. Since $\mathfrak{A}$ depends on parameters of the potential $V(x)$, a change of the potential distribution, i.e., of the shape of the potential well (e.g., by variation of concentration), can result in $\mathfrak{A} = 0$.

Substituting the first two terms of expansion (1.33) into Eq. (1.44), we obtain, after a tedious but rather simple integration

$$I = A\mathfrak{A}^{-1}\mathfrak{G}(y)[\hbar\omega - \hbar\omega_0(0) - e\varphi]^{5/2} \tag{1.46}$$

where A is the same constant as in Eq. (1.32),

$$\mathfrak{G}(y) = \frac{5}{y}\{1 + \tfrac{3}{2}y^{-3/2}[y^{1/2} - (1 + y)\arctan y^{1/2}]\}$$

and

$$y \equiv \frac{2m\mathfrak{B}}{\mathfrak{A}}E_m$$

The plot of the $\mathfrak{G}(y)$ function is shown in Appendix 3. If $\mathfrak{A} \ll 1$ even relatively small values of E_m result in $y \gg 1$; in this case we obtain from Eq. (1.46)

$$I = AE_m^{3/2}\frac{5}{2m\mathfrak{B}}\left(1 - \frac{3\pi}{\kappa}y^{-1/2}\right) \tag{1.47}$$

Equation (1.47) represents a different functional dependence for the photo-emission current than the 5/2 power law; moreover, in the energy range $\mathfrak{A}/2m\mathfrak{B} < E_m < 1/2m\mathfrak{B}$, a relative increase of the photoemission current is predicted.

The condition $\mathfrak{A} = 0$, i.e., the resonance condition, can be determined from Eq. (1.43). In the most interesting region of low concentrations, the $0 < x < d$ range can be neglected in calculations of $f(p)$. Then, as can be seen from Eq. (1.43), $\mathfrak{A} = f(p)|_{p=0}$ becomes zero if $\mathcal{J}_0[(2\kappa^{-1}/\hbar)(2me\psi')^{1/2}] = 0$. Thence, using the numerical value of the zero-order Bessel function $\mathcal{J}_0$, we obtain for the resonance values of the parameters ψ' and κ^{-1} of the potential $V(x)$

$$\frac{\kappa^{-1}}{\hbar}(2me\psi')^{1/2} = 1.2 \tag{1.48}$$

For practical calculations, condition (1.48) should be rewritten in the form $\kappa^{-2}\psi' = 5.5$, where ψ' is in volts and κ^{-1} in Angstroms.

When $\psi' < 0$, the argument of the $\mathcal{J}_0$ function becomes purely imaginary and $\mathcal{J}_0$ does not become zero for real ψ' and κ^{-1} values. Therefore the resonance effect is absent if $\psi' < 0$.

When the values of the parameters of the diffuse part of the double layer are close to those satisfying condition (1.48), the photocurrent $I(\omega)$ is given by Eq. (1.47) (at a given fixed value of φ). The resonance increase of the photocurrent may also occur at fixed ω and φ values, as shown by the dependence of I on the electrolyte concentration c_{el}. At a concentration c_{el}^*, corresponding to the resonance value of $\kappa^{-2}\psi'$, the current I should pass through a maximum.

Thick Adsorption Layers

We shall now discuss laws of photoemission which apply in the case of adsorption, for example, of organic molecules with a long (10 Å and more) hydrocarbon chain at an electrode. At high coverages, the adsorbed layer is then of sufficient thickness d_{ad}.

The general character of the potential distribution in the absence of external fields and charges is shown in Fig. 1.3. It should be mentioned that the effective permittivity ε_{ad} of the adsorbed layer within $0 < x < d_{ad}$ is usually of the same order of magnitude as that of the compact double layer (which occupies now the $d_{ad} < x < (d_{ad} + d)$ region, not shown in Fig. 1.3). Therefore, when $d_{ad} \gg d$, virtually the whole potential drop occurs over the adsorbed layer, and the range $[d_{ad}; d_{ad} + d]$ can be neglected. The potential distribution $V(x)$ from Eq. (1.7), can be then represented by

$$V(x) = \begin{cases} -e\varphi(1 - x/d_{ad}) - e^2/4\varepsilon_{ad}x + U(x), & 0 \leqslant x \leqslant d_{ad} \\ 0, & x > d_{ad} \end{cases} \quad (1.49)$$

where the applied potential φ is measured with respect to the potential of zero charge. The second term in the first line of Eq. (1.49) represents the image forces acting on the emitted electron in the range $0 < x < d_{ad}$, similar to those arising during emission into dielectrics. The $U(x)$ potential takes into account the difference in structural properties of the media occupying the regions $0 < x < d_{ad}$ and $x > d_{ad}$ (i.e., the adsorbed layer and

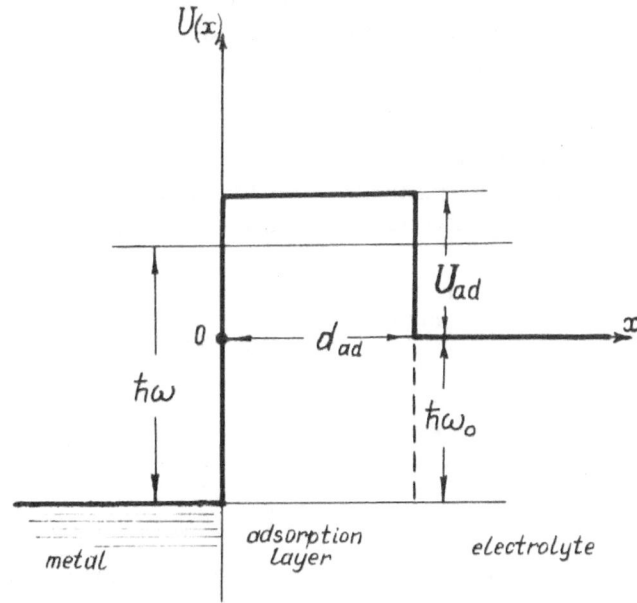

Fig. 1.3. The potential barrier at the interface — the case of a thick adsorbed layer.

bulk solution). They differ primarily in their electronic polarizability (or, more precisely, in the values of electronic work functions). The dipole potential at the adsorbate–solution interface contributes to the value of $U(x)$ as well. It can be assumed that approximately $U(x) = U_{ad} = \text{const}$, where U_{ad} is a certain averaged energy parameter which characterizes, in the "three-layered" system in question (Fig. 1.3), the region occupied by the adsorbed substance.

Let us assume that U_{ad} and d_{ad} satisfy the inequality

$$2mU_{ad}d_{ad}^2/\hbar^2 \gg 1$$

For example, when $d_{ad} \simeq 20$ Å it is enough that $U_{ad} \gtrsim 0.1$ eV. Also let it be assumed that the light frequency ω satisfies the expression $\hbar(\omega - \omega_0) < U_{ad}$; in this case, the electrons must tunnel through a wide potential barrier. In view of the assumptions previously discussed, the electron motion can be described in this case by the quasi-classical approximation. It follows from Eqs. (1.7), (1.12), and (1.14) that the photocurrent density is given by

$$I = I_{ad} \exp(-b\varphi) \tag{1.50}$$

where I_{ad} and b are potential-independent quantities expressed in tems of the characteristic parameters of the potential $V(x)$.

Equation (1.50) predicts that under the conditions described above, a significant deviation from the 5/2 power law should be observed, particularly with respect to a decrease of photocurrent, and to the voltametric curve, which should be linear in an $\ln I$ vs. φ coordinate system. Using the experimental values of b and I_{ad}, together with values of A and $E_m^0 \equiv \hbar\omega - \hbar\omega_0(0)$, characteristic of the photoemission current from the same surface in the absence of adsorption [cf. Eq. (1.32)], we can determine the parameters of the adsorbed layer. In particular, for the potential model discussed [Eq. (1.49)], we obtain (after tedious but not complex calculations) the approximate expression

$$d_{ad} = \left(\frac{b}{e}\right)^{1/2} \left[\ln\left(\frac{Ae^2(E_m^0)^{1/2}}{b^2 I_{ad}}\right)\right]^{1/2} \frac{\hbar}{(2m)^{1/2}} F_d(v)$$

$$U_{ad} = E_m^0 + \frac{1}{4}\frac{e}{b} \ln\left(\frac{Ae^2(E_m^0)^{1/2}}{b^2 I_{ad}}\right) F_U(v) \tag{1.51}$$

$$v \equiv \varepsilon_{ad} \frac{\hbar}{e^2(2m)^{1/2}} \left(\frac{e}{b}\right)^{1/2} \ln\left(\frac{Ae^2(E_m^0)^{1/2}}{b^2 I_{ad}}\right)$$

For calculations of E_m^0 and b the φ potential is measured with respect to the potential of zero charge in the absence of specific adsorption.

Plots of the functions $F_d(v)$ and $F_U(v)$ are shown in Appendix 4. The analytical expressions can be found in a paper by Brodskii *et al.* (1971). However, Eqs. (1.50) and (1.51) differ somewhat from similar expressions in Brodskii's paper and are more convenient for treatment of experimental data.

Thus, the model approach to the effect of double-layer structure on photoemission shows that comparison of experimental and calculated data can supply numerical values of the model parameters characterizing the interface.

It should be mentioned that a number of model calculations connected with the effect of adsorption of relatively small (as compared to λ) particles, including the effect of nonhomogeneous potential caused by adsorption, are presented in the paper by Brodskii *et al.* (1971) quoted above.

1.6. Dependence of the Photoemission Current on the Characteristics of the Irradiation

Apart from the dependence of the photocurrent I on the difference $\omega - \omega_0$ discussed earlier, another, weaker dependence of I on light frequency ω, as well as on polarization and intensity of the light, will now be considered. This requires a more detailed discussion of the properties of the dimensionless quantity $|\Lambda|^2$ which appears in Eqs. (1.12)–(1.14); the threshold approach is insufficient for this purpose.

A systematic quantum-mechanical calculation of excitation probability was carried out only for models of noninteracting electrons in "potential boxes" of various forms (Adawi, 1964; Bunkin and Fedorov, 1965; Schaich and Ashcroft, 1970). It can be easily shown, using the perturbation theory of electromagnetic interactions, that $|\Lambda|^2 \propto |M^{(n)}|^2$, where $M^{(n)}$ is a matrix element of the n-photon photoeffect, i.e., of the electron transfer between the initial and final states, upon absorption of n quanta. For single-photon transfer

$$M^{(1)} = \int dx\, dy\, dz\, \chi_f(x, y, z) V_\omega \psi_i(x, y, z) \tag{1.52}$$

where ψ_i and χ_f are solutions of the Schrödinger equation for the whole space (i.e., inside and outside the metal), with the initial and final energies E_i and $E_f = E_i + \hbar\omega$, respectively, and V_ω is the operator describing the electromagnetic interaction in the first-order perturbation theory, which depends on the intensity of the electric vector of the light wave. Outside the metal, the function ψ_i is quickly attenuated and χ_f satisfies an equation of the type (1.6); integration is carried out over the whole space.

Without going into the detailed form of the functions ψ_i and χ_f, it can be shown that the frequency dependence of $M^{(1)}$ is fully determined in the near-threshold energy region by the behavior of the electromagnetic field (Brodskii and Gurevich, 1973).

Consider first the case of a single-photon surface photoeffect. The x component of the intensity of the light wave field, $\mathscr{E}_x(x)$, is the determining factor in this case. For linearly polarized light the amplitude of the x field component of the incident plane wave is given by

$$\mathscr{E}_x^{(0)} = \mathscr{E}^{(0)} \sin \vartheta \cos \Delta$$

where $\mathscr{E}^{(0)}$ is the amplitude of the wave field, ϑ is the angle of incidence, and Δ is the angle between the plane of incidence and the plane of vibrations of the electric vector of the polarized wave. Regardless of its dependence on the character of reflection of the light wave from the surface and the behavior of the latter in the metal, the quantity $\mathscr{E}_x(x)$ should be proportional to $\mathscr{E}^{(0)} \cos \Delta$. Assuming that the electromagnetic field freely penetrates into the metal, we have $\mathscr{E}_x(x) \simeq \mathscr{E}_x^{(0)}$. Direct calculation of the matrix element (1.52), using the dipole approximation, results in this case in

$$I \propto \omega^{-4} \sin^2 \vartheta \cos^2 \Delta \tag{1.53}$$

Another approximation consists in utilizating Fresnel's formulas describing the behavior of $\mathscr{E}_x(x)$ in terms of macroscopic electrodynamics (Landau and Lifshitz, 1959; Born and Wolf, 1964). The metal is assumed to be characterized by a complex permittivity $\varepsilon(\omega)$, while outside the metal $\varepsilon = 1$, i.e., the dielectric constant is discontinuous at $x = 0$. The resulting photoemission current is given under conditions of surface excitation by (Feibelman, 1974)

$$I \propto \omega^{-4} \cos^2 \Delta \sin^2 \vartheta \left[\frac{2 \cos \vartheta}{\varepsilon(\omega) \cos \vartheta + [\varepsilon(\omega) - \sin^2 \vartheta]^{1/2}} \right]^2 \tag{1.54}$$

For $\varepsilon(\omega) = 1$, Eq. (1.54) reduces to (1.53). The frequency dependence is determined now by ω^{-4} and the $\varepsilon(\omega)$ term. It should be mentioned, however, that the validity of the direct substitution into the microscopic relations for matrix elements [of the type (1.52)] of the value of $\mathscr{E}_x(x)$, found assuming discontinuity of $\varepsilon(x)$, is to be seriously questioned.

Therefore it seems reasonable to present another model which can be used for describing the dependence of the field $\mathscr{E}_x$ on x. The permittivity of metals, $\varepsilon(\omega)$, has, within the frequency range discussed, a negative real part Re $\varepsilon < 0$. Since the permittivity outside the metal is real and positive (i.e., the external medium does not absorb light), the quantity Re $\varepsilon(\omega)$ (being a function of x) must pass through zero at some point $x = x_\omega$ near the surface. The field strength $\mathscr{E}_x(x)$ has a sharp maximum in the vicinity of x_ω, and, therefore, the latter point may be the most important one with respect to the surface photoexcitation.

The dependence of $\mathscr{E}_x$ on x is given in this case by (Landau and Lifshitz, 1959)

$$\mathscr{E}_x(x) \propto \frac{\mathscr{E}^{(0)} \cos \Delta}{a(x - x_\omega) + i \operatorname{Im} \varepsilon(\omega)} \tag{1.55}$$

where $a > 0$ depend to a slight extent on ω; $\operatorname{Im} \varepsilon$ $(0 < \operatorname{Im} \varepsilon \ll 1)$ denotes the imaginary part of $\varepsilon(\omega)$.

Calculations based on the above model result (Brodskii and Gurevich, 1973) in

$$I \propto \omega^{-2} \cos^2 \Delta \tag{1.56}$$

In the case of bulk photoexcitation, the photocurrent is determined by the field of the light wave which penetrates into the metal. If the field inside the metal varies relatively slowly with distance so that its changes are negligible within the thickness of the photoemitting layer, the following approximation holds:

$$I \propto \omega^{-4}[1 - \mathscr{R}(\omega, \vartheta, \Delta)]g(\vartheta, \Delta) \tag{1.57}$$

where $\mathscr{R} > 0$ is the reflection coefficient (its dependence on frequency and angles is determined from the Fresnel formulas). The frequency factor ω^{-4} is given by the dependence on ω of the quantity $M^{(1)}$. Finally the factor $g(\vartheta, \Delta)$ describes the dependence on ϑ and Δ of the photoexcitation probability. Generally, the relation $g(\vartheta, \Delta)$ depends on the crystal structure and on orientation of the crystallographic axes relative to the interface. However, the relation $g(\vartheta, \Delta)$ is often assumed to be constant and the angular variation of the photoemission current is determined by the dependence of the reflection coefficient $\mathscr{R}$ on ϑ and Δ.

The dependence of the photoemission current $I^{(n)}$ (or of the quantum yield $Y^{(n)}$) of the n-photon photoeffect on the parameters discussed is calculated (in terms of the models used) in a similar way. In particular, the dependence of $Y^{(2)}$ on ϑ was described by Barashev (1970) for the case of the surface photoeffect in terms of a discontinuous change of optical constants at the interface.

The dependence of the photocurrent on the intensity of the incident light J_0 or on the absorbed light J (proportional to J_0) is given in terms of the perturbation theory by

$$I^{(n)} \propto J_0{}^n \tag{1.58}$$

regardless of the character of the photoexcitation.

It should be mentioned, however, that in the field of a light wave of high intensity ($\mathscr{E}^{(0)} \gtrsim 10^6$ V/cm), deviations from the simple relation (1.58) should be observed, due to the appearance of nonlinear effects.

Theory of Photodiffusion Currents

2.1. Formulation of the Problem and Basic Relations

It was already mentioned in Section 0.2 that "primary" electrons emitted into solution undergo further transitions in the latter phase. The magnitude of the measured photocurrent depends considerably on the nature of these transitions. Therefore, the physical nature and the characteristic behavior of "excess" electrons in solution play a very important part in the process discussed. (The "excess" here refers to the number of electrons in atomic shells, which is unambiguously determined by the total charge of nuclei in the atoms of solutes and solvent.)

A detailed discussion of these problems, essential in many fields of radiation physics and physical chemistry, is beyond the scope of this monograph, the more so because many questions still remain unsolved. The scope of the discussion which follows is limited to a short qualitative presentation of modern views concerning the behavior of excess electrons in liquids (Hart and Anbar, 1970; Jortner and Kestner, 1973; Pikaev, 1969) necessary for an understanding of the text which follows.

It is assumed on the basis of a large body of experimental material, as well as on model-based theoretical calculations, that excess electrons in polar media (e.g., aqueous electrolyte solutions) exist in two qualitatively different states. Electrons of the first type (delocalized) may be direct carriers of electrical current, as in the case of conduction electrons in solids. Electrons of the second type (localized) can carry current only by joint motion with localization centers or by jumps. The mobility of delocalized electrons is obviously very much higher than that of localized electrons. Delocalized electrons are called "dry" or quasi-free, whereas electrons in bound states are called solvated (in water, "hydrated"), according to the terminology accepted in the fields of radiation chemistry and physics of polar liquids.

Early papers assumed that solvated electrons appeared owing to the orientation polarization of the medium, in analogy to polarons in solids (Davydov, 1948; Dejgen, 1954). This model was later modified (Jortner, 1959) by taking into account formation of cavities in the medium of solvent molecules. Accumulation of new experimental data has led recently to a large number of further theoretical studies.

On the other hand, model-based calculations on particles in a stray field (Mott and Davis, 1971) show that an energy level exists (similar to the lower edge of the conduction band in ordered structures) separating the higher energy region in which delocalized states arise from the lower one where localized states appear. Such localization, derived on the basis of a single-particle model, differs qualitatively from that discussed above. In a real polar liquid, however, an electron which has become localized by collective interactions in a "stray" configuration of an external (i.e., with respect to the electron) potential can be transformed into a "common" solvated electron.

It should also be noted that the calculations mentioned above provide some basis for the model of the external (with respect to the metal) medium described in Chapter 1 for the act of photoemission. Effects which arise on account of interaction of emitted electrons with this medium were treated there by introducing a certain constant potential determined by the properties of the medium and by ascribing an effective mass to electrons.

No final conclusions can be drawn at present concerning the detailed physical nature of excess electrons in electrolytic solutions. Several quantitative characteristics of their behavior are also lacking. Nevertheless, photoelectron emission into solutions can still be investigated, since the most important fact for these studies consists in the very existence of delocalized ("dry") and solvated electrons.

The complete scheme of physicochemical transformations following photoemission should obviously include not only thermalization and solvation, but also the direct interaction of the "dry" (and possibly also "hot") electron with acceptors present in solution. A systematic mathematical description of motion and inelastic interaction of "dry" electrons in solution leads to serious difficulties. In particular, the use of diffusion equations requires, generally speaking, justification. However, over a wide range of variation of experimental parameters, "dry" electrons need not be taken into account. Lam and Hunt (1975), Bronskill *et al.* (1970), and Aldrich *et al.* (1971) demonstrated that the effect of a direct interaction of acceptors with "dry" electrons becomes noticeable only for high acceptor concentrations (mole fraction not too small compared with unity). Even the most effective acceptors of "dry" electrons (Cd^{2+}, NO_3^-, acetone) compete successfully with water in capturing "dry" electrons only at concentrations of the order of 1 mole/liter. Photoemission studies can be successfully carried out at much lower acceptor concentrations (10^{-3}–10^{-1} moles/liter). The experiments described below were carried out in this concentration range, except for the acceptor Cd^{2+} (which according to Hart *et al.* virtually does not capture "dry" electrons) and NO_3^- (cf. Section 8.1). The contribution of the process in which "dry" electrons are returned to the electrode was shown by special experiments (cf. Section 5.2) to be negligibly small in the systems considered.

Thus, conditions exist under which electrons emitted into the solution thermalize and become solvated without previous interactions. After solvation they can diffuse in the bulk and back to the emitting electrode; also, they may participate in homogeneous reactions with electron acceptors A and be captured by the electrode surface (cf. Fig. 0.1). From the thermodynamic point of view, the capture of "dry" and solvated electrons by the electrode is energetically advantageous. In fact, the potential range usually covered in photoemission studies is much more positive than the equilibrium potential of the "electron" electrode, which corresponds to the actual concentration of solvated electrons in solution. If the electrode potential is sufficiently close to the equilibrium potential of the "electron" electrode, electron transfer from the metal into solution is possible even in the absence of illumination (cf. Section 10.3). Moreover, the capture of electrons by the electrode surface can be kinetically inhibited. This is reflected in the finite value of the rate constant introduced below.

Taking into account possible electrode reactions involving products of electron capture by acceptors [eA], we can express the experimentally observed photocurrent as a superposition of separate currents

$$j = I - I_e \pm I_{eA} \tag{2.1}$$

where I is the emission current, I_e is the return current of solvated electrons, and I_{eA} is the current of the electrode reaction involving [eA].

If other possible transformations of [eA] products and related electrode reactions need to be considered, Eq. (2.1) must be modified [see, e.g., Eq. (2.21)].

The processes discussed occur at distances much smaller than the dimensions of an electrochemical cell. Therefore the emitting electrode can be considered to be flat and to occupy the half-space $x < 0$; the counter-electrode is placed at infinity. Correspondingly, currents of solvated electrons and [eA] products transported to the counter-electrode are negligibly small. It follows from the law of charge conservation that, in this case, the number of electrons captured by acceptors in the bulk equals the difference between the number of emitted electrons and the number of electrons which are returned to the emitting electrode. Equation (2.1) can be rewritten for stationary conditions in the form

$$j = \nu(I - I_e) \tag{2.2}$$

where the coefficient ν is analogous to the stoichiometric number; ν equals the total number of electrons crossing the interface per one electron emitted. If the product [eA] undergoes one electron oxidation at the electrode,

$\nu = 0$. (Also $j = 0$ since all emitted electrons revert to the electrode.) If the [eA] products are not involved in electrode reactions, $\nu = 1$ [$I_{eA} = 0$, cf. Eq. (2.1)]. Finally for one-electron reduction of [eA], $\nu = 2$. In principle, fractional values of the stoichiometric coefficient are possible in the case of parallel electrode reactions (cf. Chapters 7 and 8).

It is clear from physical considerations (as well as from the calculations given below) that under conditions of high acceptor concentration, virtually all electrons are captured by acceptors and the reverse current is absent, i.e., $I_e = 0$ and $j = \nu I$. It can be seen from Eq. (2.2) that the photocurrent observed in this case is very simply connected with the photoemission current, providing a possibility for the experimental investigation of I.

The characteristic distances for processes connected with motion and chemical reactions of solvated electrons and their capture by acceptors considerably exceed interatomic distances. Therefore diffusion-type equations should be used in the quantitative description of these processes accompanying photoelectron emission, and the currents connected with such processes are called photodiffusion currents (Gurevich and Rotenberg, 1968).

It follows from Eq. (2.2) that the description of photodiffusion phenomena requires, in the simplest cases, calculation only of the "reverse" electron current I_e. If emission proceeds in a highly concentrated solution and the electric field in the diffuse layer can be neglected, the concentration distribution of solvated electrons c_e in the bulk solution is given by

$$\frac{\partial c_e}{\partial t} = \mathscr{D}_e \frac{\partial^2 c_e}{\partial x^2} - k_A c_A c_e + \Phi(x, t) \tag{2.3}$$

where $\mathscr{D}_e$ is the diffusion coefficient of solvated electrons, k_A is the rate constant of the acceptor–solvated electron reaction, and $c_A(x)$ is the bulk concentration of the acceptor (generally x dependent).

It is assumed that there is a sufficient number of acceptors in solution and their concentration c_A remains constant in spite of interactions with electrons. Equation (2.3) is derived assuming that the effect of the electric field created by solvated electrons themselves can be neglected. It is valid, anyway, if $\kappa^{-3}c_e \ll 1$ (κ^{-1} designates here the screening length in solution, cf. p. 30; c_e is the number of electrons per cubic centimeter).

The function $\Phi(x, t)$ in Eq. (2.3) describes the rate of generation of solvated electrons, per unit time, from "primary" electrons emitted into solution. The general form of this function is determined by the initial energy distribution of electrons and by the nature of their retardation and solvation in the electrolyte. Therefore, it depends parametrically on the mean energy of emitted electrons. Calculations of $\Phi(x, t)$ are based on various models which, in principle, can be experimentally verified (cf. Section 5.3).

If the contribution associated with direct capture of "dry" electrons by acceptors is negligibly small, then, regardless of the actual form of the function $\Phi(x, t)$, the following relation (derived from the law of charge conservation) applies:

$$e \int_0^t dt \int_0^\infty \Phi(x, t) \, dx = \int_0^t I(\tau) \, d\tau$$

where $t = 0$ corresponds to the initiation of the experiment [for $t < 0$, $\Phi(x, t) \equiv 0$]. Under stationary conditions, the above expression reduces to

$$\int_0^\infty \Phi(x) \, dx = I/e \qquad (2.4)$$

The boundary conditions for Eq. (2.3) are

$$c_e(\infty) = 0, \qquad k_s c_e(0) = \mathscr{D}_e \left(\frac{dc_e}{dx}\right)_{x=0} \qquad (2.5)$$

where k_s is the rate constant of the process of solvated electron capture by the electrode surface. The second boundary condition in (2.5) assumes the latter capture process to be a first-order reaction. When the rate of this reaction increases to infinity ($k_s \rightarrow \infty$), the boundary condition reduces to

$$c_e(0) = 0$$

The initial condition for nonstationary processes is obviously

$$c_e(x, 0) = 0$$

When an electric potential gradient $\mathscr{V}(x)$ exists in the solution which causes migration of solvated electrons (in particular, the potential gradient in the diffuse part of the double layer), Eq. (2.3) and the second boundary condition in (2.5) must be modified: the quantity $\mathscr{D}_e \, dc_e/dx$, which describes the diffusion flux of solvated electrons, must be replaced by

$$\mathscr{D}_e \left[\frac{dc_e}{dx} - \frac{e}{kT} c_e(x) \frac{d\mathscr{V}}{dx}\right]$$

The second term in this expression represents the flux of migrating solvated electrons, which takes into account the relation between the mobility and the diffusion coefficient.

The concentration distribution c' of the capture products [eA] is given by

$$\frac{\partial c'}{\partial t} = \mathscr{D}' \frac{d^2 c'}{dx^2} + k_A c_A c_e \qquad (2.3a)$$

with the following boundary and initial conditions:

$$c'(\infty) = 0, \qquad \mathcal{D}'\left(\frac{\partial c'}{\partial x}\right)_{x=0} = k_{eA}c'(0), \qquad c'(x, 0) = 0$$

where $\mathcal{D}'$ is the diffusion coefficient of [eA] products and k_{eA} is the rate constant of their reaction at the electrode.

2.2. Stationary Photodiffusion Currents

General Relations

Consider stationary conditions or conditions in which there is hardly any change with time. From (2.3)

$$\mathcal{D}_e \frac{d^2 c_e}{dx^2} - k_A c_A c_e + \Phi(x) = 0 \tag{2.6}$$

Solutions of nonhomogeneous differential equations similar to (2.6) can be obtained using the following method (Korn and Korn, 1961). Assume the homogeneous equation

$$\frac{d^2 c}{dx^2} - S(x)c = 0 \tag{2.7}$$

[where $S(x)$ is a function independent of c and $S(x) > 0$ for $x \to \infty$] to have two known, linearly independent solutions c_1 and c_2. One of these solutions, e.g., c_1, can be chosen so that $c_1(\infty) = 0$. Since the Wronskian $W[c_1, c_2] \equiv c_1(dc_2/dx) - c_2(dc_1/dx)$ is independent of x, normalization of this solution can be chosen so that $W[c_1, c_2] = 1$. The solution $c_e(x)$ (which equals zero for $x \to \infty$) of the nonhomogeneous equation (2.6) is given by

$$c_e(x) = c_1 \int_0^x c_2 \frac{\Phi(x)}{\mathcal{D}_e} dx + c_2 \int_x^\infty c_1 \frac{\Phi(x)}{\mathcal{D}_e} dx \tag{2.8}$$

The unknown constant included in c_2 is determined from the boundary condition for $x = 0$. In particular, if the boundary condition is in the form of (2.5) and the "reverse" current of solvated electrons, I_e, is given by $I_e = e\mathcal{D}_e(dc_e/dx)|_{x=0}$, we obtain from Eqs. (2.8), (2.5), and the condition $W[c_1, c_2] = 1$,

$$I = e\left[c_1(0) - \frac{\mathcal{D}_e}{k_s}\left(\frac{dc_1}{dx}\right)_{x=0}\right]^{-1} \int_0^\infty c_1(x)\Phi(x)\, dx \tag{2.9}$$

Examples of results obtained using Eqs. (2.8) and (2.9) are discussed below.

Photocurrent in the Absence of External Fields

In the simplest case, $S(x) = \text{const} = Q^2$, where $Q = (k_A c_A / \mathscr{D}_e)^{1/2}$. Solution c_1 of Eq. (2.7) is then of the form $c_1 = ae^{-Qx}$ (a is a constant). Whence, using Eq. (2.9), we obtain

$$I_e = \frac{e}{1 + Q\mathscr{D}_e/k_s} \int_0^\infty \Phi(x) e^{-Qx} \, dx \qquad (2.10)$$

Substitution of Eq. (2.10) into Eq. (2.2) results in the final expression

$$j = ev \int_0^\infty \Phi(x) \left[1 - \frac{1}{1 + Q\mathscr{D}_e/k_s} e^{-Qx} \right] dx \qquad (2.11)$$

Equation (2.11) describes simultaneously the dependence of the photocurrent j on the form of the function $\Phi(x)$ and on Q, i.e., on the concentration of electron acceptors c_A. In limiting cases of high and low acceptor concentration, an explicit dependence of j on Q can be obtained regardless of the form of the function $\Phi(x)$.

For high acceptor concentrations, when $Qx_* \gg 1$ (x_* designates the size of the region in which $\Phi(x)$ differs considerably from zero) the exponential in Eq. (2.11) can be neglected and

$$j = vI \qquad (2.12)$$

In this case, virtually all emitted electrons are captured by acceptors in solution.

At sufficiently low concentrations, when $Qx_* \ll 1$, the exponential term in Eq. (2.11) can be expanded. Using the first two terms of the expansion we obtain

$$j = v \frac{x_0 + \mathscr{D}_e/k_s}{1 + Q\mathscr{D}_e/k_s} QI \qquad (2.13)$$

where x_0 is determined by

$$x_0 \equiv e/I \int_0^\infty x\Phi(x) \, dx$$

being thus the mean distance from the electrode at which the cloud of solvated electrons is formed; it is called the mean length of solvation. Obviously, x_0 is of the order of x_*.

In limiting cases, Eq. (2.13) reduces to the following:

(1) Electron capture by the electrode surface is sufficiently fast, so that $\mathscr{D}_e/(x_0 k_s) \ll 1$, $Q\mathscr{D}_e/k_s \ll 1$; then

$$j = vx_0 QI \qquad (2.14)$$

(2) Electron capture at the electrode occurs at a finite rate (x_0 and $\mathscr{D}_e/k_s$ are of the same order of magnitude), but the acceptor concentration is very low ($Q\mathscr{D}_e/k_s \ll 1$); then

$$j = \nu(x_0 + \mathscr{D}_e/k_s)QI \qquad (2.15)$$

It should be mentioned that the two types of dependence of photocurrent on concentration, described by Eqs. (2.14) and (2.15), differ only by constant factors. In both cases the measured current is proportional to Q and, consequently, to the square root of acceptor concentration, c_A.

(3) Electrons are not captured by the electrode surface ($k_s \to 0$); then

$$j = \nu I$$

The last relation is sufficiently clear: In the absence of capture by the electrode surface, all electrons are captured by acceptors, as is the case in the limit $c_A \to \infty$. Thus, the photocurrent can attain the limiting value (νI) in two cases: (a) at high acceptor concentrations and (b) by the electrode surface at a very low capture rate of the solvated electron.

The Effect of the Diffuse-Layer Field

In dilute electrolytes, the field of the diffusion layer can significantly affect the motion of electrons in the vicinity of the electrode. The effect is closely connected with the dynamic ψ' effect for ions, described in detail in a number of papers (Matsuda and Delahay, 1960; Rangarajan, 1963; Malev, 1970).

The qualitative effect of the diffuse-layer field on photodiffusion currents was first described by Barker et al. (1966) while a quantitative description was later attempted by Bomchil et al. (1970). In the latter paper, the problem was solved numerically taking also into account the effect of diffuse-layer field on the concentration distribution of charged acceptors such as hydrogen ions.

An analytical solution of the problem was obtained by Rotenberg and Gurevich (1973) for the case of uncharged acceptors for the case where the diffuse layer affects only electron migration. The latter treatment is followed below.

In the presence of an electric field, the expression for c_e under stationary conditions is given by

$$\mathscr{D}_e \frac{d}{dx}\left[\frac{dc_e}{dx} - c_e \frac{d\overline{\mathscr{V}}}{dx}\right] - k_A c_A c_e + \Phi(x) = 0$$

where $\overline{\mathscr{V}}(x)$ is the electric potential in the diffuse layer expressed in kT/e units. It is assumed that this potential decreases exponentially, so that $\overline{\mathscr{V}}(x) = \bar{\psi}' \exp(-x/L)$, where $\bar{\psi}'$ is the ψ' potential measured in kT/e units and L is the effective length of the potential drop. The value of L is chosen so that the latter expression becomes very close for $x \simeq \kappa^{-1}$ (κ^{-1} is the screening length, cf. p. 30) to the well-known and more complex expression of Delahay (1965). The effect of the diffuse-layer field on migration of solvated electrons is significant only for $\kappa^{-1}, L \gg d$ (where d is the compact-layer thickness). Therefore it can be assumed that $(x - d)/L \simeq x/L$ [cf. Eq. (1.40)], in which $\kappa = \kappa_*$ is chosen so that Eq. (1.40) coincides with the exact expression when $x = d$. If $|\bar{\psi}'| \leqslant 1$, $L = \kappa^{-1}$ and then $V(x) = e\mathscr{V}(x)$.

Changing variables $c_e = u(x) \exp\{\frac{1}{2}\bar{\psi}' \exp(-x/L)\}$, we obtain, after simple transformations,

$$u'' - \left(\frac{1}{2}\bar{\psi}' \exp\left(-\frac{x}{L}\right) + \frac{1}{4}(\bar{\psi}')^2 \exp\left(-\frac{2x}{L}\right) + \mathfrak{M}^2\right)u$$

$$= -\frac{\Phi(x)}{\mathscr{D}_e} L^2 \exp\left\{-\frac{1}{2}\bar{\psi}' \exp\left(-\frac{x}{L}\right)\right\} \quad (2.16)$$

where $\mathfrak{M}^2 = k_A c_A L^2 / \mathscr{D}_e$. The left side of Eq. (2.16) is related to equations of the type (2.7). Assuming that the electron is captured sufficiently quickly by the electrode surface and $c_e(0) = 0$, we obtain, using Eq. (2.9),

$$I_e = \frac{\exp\left(\frac{1}{2}\bar{\psi}'\right)}{u_1(0)} e \int_0^\infty u_1(x) \Phi(x) \exp(-\frac{1}{2}\bar{\psi}' e^{-x/L}) \, dx \quad (2.17)$$

where $u_1(x)$ is the linearly independent solution of the homogeneous part of Eq. (2.16) which tends to zero as $x \to \infty$. A more general case (without limitations concerning the rate of electron capture by the surface) can be treated similarly.

The solution $u_1(x)$ is expressed by a confluent hypergeometric function $_1F_1$ (Korn and Korn, 1961):

$$u_1\left(\frac{x}{L}\right) = \exp\left(\frac{x}{2L}\right) \left[\bar{\psi}' \exp\left(-\frac{x}{L}\right)\right]^{1/2 + \mathfrak{M}}$$

$$\times \exp\left[-\frac{1}{2}\bar{\psi}' \exp\left(-\frac{x}{L}\right)\right] {}_1F_1(1 + \mathfrak{M}, 2\mathfrak{M} + 1, \bar{\psi}' e^{-x/L})$$

Substituting this expression in Eq. (2.17), the final expression is obtained as

$$j = \nu\left\{I - \frac{e}{{}_1F_1(\mathfrak{M}, 2\mathfrak{M} + 1, -\bar{\psi}')} \int_0^\infty \Phi(x) e^{-Qx}\right.$$

$$\left. \times {}_1F_1(\mathfrak{M}, 2\mathfrak{M} + 1, -\bar{\psi}' e^{-x/L}) \, dx\right\} \quad (2.18)$$

When $\psi' = 0$, $_1F_1(\mathfrak{M}, 2\mathfrak{M} + 1, 0) = I$ and Eq. (2.18) coincides with Eq. (2.11) in the limit $k_s \to \infty$. The field effect strongly decreases if $L \gg x_0$, i.e., when the thickness of the diffuse layer considerably exceeds x_*. In this case, in the integration range (determined by x_*) the function

$$_1F_1(\mathfrak{M}, 2\mathfrak{M} + 1, -\bar{\psi}'e^{-x/L})$$

becomes virtually constant and it can be assumed equal to

$$_1F_1(\mathfrak{M}, 2\mathfrak{M} + 1, -\bar{\psi}')$$

It can be easily shown, using the asymptotic expression for

$$_1F_1(\mathfrak{M}, 2\mathfrak{M} + 1, -\bar{\psi}'e^{x/L})$$

in the range $|\psi'| \gg 1$, that the flux j is either zero ($\psi' > 0$) or reaches saturation ($\psi' < 0$). This result is fully understandable from the physical point of view.

In the absence of acceptors ($c_A = 0$), the photocurrent is formally equal to zero, i.e., all electrons are returned to the electrode, regardless of electrostatic repulsion ($\psi' < 0$). However, it should be remembered that the condition assumed, $c_e(\infty) = 0$, cannot, strictly speaking, be satisfied under stationary conditions if $c_A = 0$.

Calculation of the photocurrent [Eq. (2.18)] requires the form of the function $\Phi(x)$ to be known. Rotenberg and Gurevich (1973) numerically integrated the photocurrent as a function of electrolyte concentration for a few simple models. The results of their calculations are discussed in detail in Section 5.3.

Photodiffusion Currents under Conditions of Acceptor Discharge

If acceptors undergo cathodic reduction, their concentration in the vicinity of the electrode decreases and becomes zero under limiting current conditions. In this case

$$\mathscr{D}_e \frac{d^2c_e}{dx^2} - k_A c_A(x)c_e + \Phi(x) = 0 \qquad (2.19)$$

If the acceptor concentration $c_A(x)$ is assumed to depend linearly on the distance from the electrode within the diffusion layer δ_N (Nernst model), then

$$c_A(x) = \begin{cases} \dfrac{c_A{}^0}{\delta_N} x & \text{for} \quad x < \delta_N \\[2ex] c_A{}^0 & \text{for} \quad x > \delta_N \end{cases}$$

where $c_A{}^0$ is the acceptor concentration in the bulk solution. Actually, the concentration profile near the external edge of the diffusion layer, i.e., for

$x \simeq \delta_N$, differs from the Nernst model (cf., e.g., Pleskov and Filinovskii, 1975). However, the linear concentration profile can still be considered valid for the range of interest x, since the distance x_0 at which solvated electrons are formed can be much less than δ. The partial solution of the corresponding homogeneous equation (equal to zero at infinity) is

$$c_1(x) = \sqrt{\frac{\pi \xi}{3}} \{I_{-1/3}(\tfrac{2}{3}\xi^{3/2}) - I_{1/3}(\tfrac{2}{3}\xi^{3/2})\}$$

where $\xi \equiv (c_A{}^0 k_A / \mathscr{D}_e \delta_N)^{1/3} x$ and $I_q(z)$ is a modified Bessel function of the qth order. In region x, where $\Phi(x)$ significantly differs from zero, $\zeta \ll 1$ and

$$c_1(x) = \pi^{1/2} \left\{ \frac{1}{3^{2/3}\Gamma(2/3)} - \frac{x}{3^{4/3}\Gamma(4/3)} \right\}$$

where $\Gamma(z)$ is the gamma function.

Finally, the photocurrent j_* corresponding to the limiting condition $c_e = 0$ is given (Rotenberg, 1973) by

$$j_* = \nu(3)^{1/3} \left(\frac{k_A c_A{}^0}{\mathscr{D}_e \delta_N} \right)^{1/3} x_0 I \tag{2.20}$$

The dependence of the photocurrent on acceptor concentration differs here from that described by Eq. (2.14). In fact, under limiting-current conditions of acceptor reduction $j_* \propto (c_A{}^0)^{1/3}$, whereas for electrochemically inactive acceptors, $j \propto (c_A{}^0)^{1/2}$.

Photodiffusion Currents under Conditions of Homogeneous Chemical Reaction

The products [eA] can also take part in homogeneous chemical reactions. Let it be assumed that the product of such reaction [R] can be oxidized or reduced, and similarly with the species [eA], at the electrode, depending on the electrode potential (cf. Fig. 0.1).

The total photocurrent due to photoemission and electrode reactions of [eA] and [R] is given in this case by

$$j = I - I_e - \nu_{eA} I_{eA} - \nu_R I_R \tag{2.21}$$

where I_{eA} and I_R are the currents for the electrochemical reactions of [eA] and [R], and ν_{eA} and ν_R are stoichiometric coefficients of these reactions, respectively. According to Eq. (2.21) $\nu_{eA,R} = 1$ for one-electron oxidation and $\nu_{eA,R} = -1$ for one electron reduction.

It follows from the law of charge conservation that under stationary conditions

$$I_{eA} + I_R = I - I_e \tag{2.22}$$

The currents I_{eA} and I_R can be obtained by solution of the system of diffusion equations for solvated electrons and products [eA]

$$\mathscr{D}_e \frac{d^2 c_e}{dx^2} - k_A c_A c_e + \Phi(x) = 0$$

$$\mathscr{D}' \frac{d^2 c'}{dx^2} - k_v c' + k_A c_A c_e = 0$$

where k_v is the rate constant of the homogeneous reaction of [eA].

The above system was formulated assuming that the products [eA] (or [R]) do not recombine in solution. The photocurrent I_e is given by Eq. (2.10), and I_{eA}, by solution of the system of equations shown above (Eletskii *et al.*, 1970):

$$I_{eA} = \frac{e}{1 + Q_v \mathscr{D}'/k_{eA}} \times \frac{Q^2}{Q^2 - Q_v^2} \int_0^\infty \Phi(x)(e^{-Q_v x} - e^{-Qx})\, dx \quad (2.23)$$

where $Q_v = (k_v/\mathscr{D}')^{1/2}$. It is easy to obtain the resulting photocurrent j by substituting Eqs. (2.23) and (2.10) into (2.21) and noting (2.22).

The general expression is rather involved, so the treatment will be restricted to a few important limiting cases:

1. Products [eA] and [R] are either both oxidized, or both reduced, within the whole potential range. The total current j then equals either $2(I - I_e)$ or 0, i.e., the kinetic relations do not differ from those discussed above.

2. Product [eA] becomes oxidized and [R] reduced at the electrode. Assuming for simplicity $Q\mathscr{D}_e/k_s \gg 1$, we have

$$j = 2\left\{ I - e \int_0^\infty \Phi(x) e^{-Qx}\, dx \right.$$
$$\left. - \frac{e}{1 + Q\mathscr{D}'/k_{eA}} \times \frac{Q^2}{Q^2 - Q_v^2} \int_0^\infty \Phi(x)[e^{-Q_v x} - e^{-Qx}]\, dx \right\} \quad (2.24)$$

For high acceptor concentrations ($Qx_0 \gg 1$) and a sufficiently low rate constant k_v (so that $Q_v \mathscr{D}'/k_{eA} \ll 1$ and $Q_v x_0 \ll 1$), Eq. (2.24) reduces to

$$j = 2x_0 Q_v I \quad (2.25)$$

Thus, the photocurrent is controlled by the rate of the homogeneous reaction of the oxidizing product [eA].

When $Qx_0 \ll 1$

$$j = 2x_0 \frac{QQ_v}{Q + Q_v} I \quad (2.26)$$

Also when $Q_v \ll Q$, Eq. (2.26) reduces to Eq. (2.25); and when $Q_v \gg Q$, it reduces to Eq. (2.14).

3. Product [eA] is reduced at the electrode and product [R] may be oxidized, or reduced, depending on potential. Making the same assumptions with respect to k_v as above, we obtain for low acceptor concentrations ($Qx_0 \ll 1$):

$$j = \frac{j_{k_v=0}}{2} \left(1 + \frac{Q}{Q + Q_v} - \nu_R \frac{Q_v}{Q + Q_v} \right) \qquad (2.27)$$

where $j_{k_v=0}$ is the photocurrent in the absence of the bulk reaction ($Q_v = 0$). The quantity ν_R in Eq. (2.27) is generally potential dependent. In the simplest case, if only products [R] are oxidized at the electrode, $\nu_R = 1$ and

$$j = j_{k_v=0} \frac{Q}{Q + Q_v} \qquad (2.28)$$

Equations (2.27) and (2.28) were derived by Barker (1968) and later generalized by Barker and Bolzan (1974a) by analysis of equivalent electric circuits.

Photodiffusion Currents in the Presence of Two Acceptors

Let electrons be emitted into a solution containing two acceptors of solvated electrons, A and B; [eA] becomes reduced and [eB] oxidized at the electrode. The stationary photocurrent j is obviously equal to twice the value of the reduction current for [eA]: $j = 2I_{eA}$. Let [eA] remain unchanged in the bulk solution. Integration of Eq. (2.2a) with respect to x between 0 and ∞ results in

$$I_{eA} = e \int_0^\infty k_A c_A c_e \, dx$$

since $\partial c'/\partial t = 0$ and the flux of [eA] is given by $-\mathscr{D}' \, dc'/dx$.

Similarly, replacing $k_A c_A$ by $k_A c_A + k_B c_B$ in Eq. (2.6), we obtain for the boundary condition $c_e(0) = 0$:

$$\int_0^\infty (k_A c_A + k_B c_B) \, dx = \int_0^\infty \Phi(x) \left\{ 1 - \exp\left[-x \left(\frac{k_A c_A + k_B c_B}{\mathscr{D}_e} \right)^{1/2} \right] \right\} dx$$

Since c_B and c_A are independent of x, the latter two expressions result in

$$j = \frac{2k_A c_A}{k_A c_A + k_B c_B} e \int_0^\infty \Phi(x) \left\{ 1 - \exp\left[-x \left(\frac{k_A c_A + k_B c_B}{\mathscr{D}_e} \right)^{1/2} \right] \right\} dx \qquad (2.29)$$

At sufficiently low concentrations of both acceptors, when

$$x_0 \left(\frac{k_A c_A + k_B c_B}{\mathscr{D}_e} \right)^{1/2} \ll 1$$

the result

$$j = 2k_A c_A x_0 I / \mathcal{D}_e^{1/2}(k_A c_A + k_B c_B) \tag{2.29a}$$

is obtained. At a sufficiently high concentration of at least one of the acceptors, we have

$$j = 2k_A c_A I / (k_A c_A + k_B c_B)$$

The above expressions allow the ratio k_A/k_B of the rate constants of electron capture to be determined from the experimental dependence of the photocurrent on concentration of one of the acceptors.

2.3. Alternating Photodiffusion Currents

The laws of photodiffusion reactions proceeding under conditions of illumination having alternating intensity will now be discussed. Owing to the inertness of the diffusion process and to the finite rate of chemical bulk and electrode reactions, the time dependences of I_e, I_{eA}, and I should differ in the general case. Investigation of these differences can provide, as shown below, additional information on the kinetics of chemical and electrochemical reactions, including those proceeding with formation of intermediates.

Periodic Time Dependence of Light Intensity

Let the light intensity vary with frequency Ω. The corresponding components of the photoemission current are then given by $\tilde{I} = I^{(0)} \exp(i\Omega t)$, where $I^{(0)}$ is the amplitude of the periodic photocurrent. If the modulation frequency Ω is lower than the reciprocal transition and solvation time (i.e., less than $10^{12} \sec^{-1}$) the source function $\tilde{\Phi}$ is given by $\tilde{\Phi}(x) = \Phi^{(0)}(x)e^{i\Omega t}$ and, as follows from Eq. (2.4),

$$\int_0^\infty \Phi^{(0)}(x) \, dx = I^{(0)}/e$$

After a sufficiently long period has elapsed from the start of the experiment so that transition processes can be considered finished, the solution of Eqs. (2.3) and (2.3a) should be sought in the forms $\tilde{c}_e = c(x)e^{i\Omega t}$ and $\tilde{c}' = c'(x)e^{i\Omega t}$.

In the simplest case, when the products [eA] are not discharged at the electrode, we obtain from Eq. (2.3) an expression of similar structure to that of Eq. (2.11) (Gurevich and Rotenberg, 1968):

$$j = \exp(i\Omega t)e \int_0^\infty \Phi^0(x)\left[1 - \frac{1}{1 + Q_\Omega \mathcal{D}_e/k_s} e^{-Q\Omega x}\right] dx \tag{2.11a}$$

where

$$Q_\Omega \equiv [(i\Omega + k_A c_A)/\mathscr{D}_e]^{1/2} = Q_1 + iQ_2$$

$$Q_1 = \frac{\Omega}{(2\mathscr{D}_e)^{1/2}} [(\Omega^2 + k_A^2 c_A^2)^{1/2} - k_A c_A]^{-1/2}$$

$$Q_2 = \frac{\Omega}{2\mathscr{D}_e Q_1}$$

Since Q_Ω is a complex number, the quantity $\tilde{j}$ is given by $\tilde{j} = j^{(0)} e^{i(\Omega t + \Theta)}$, where $j^{(0)}$ is the amplitude of the resulting photocurrent and Θ is the phase shift between $\tilde{j}$ and $\tilde{I}$ due to the diffusion processes. It follows from Eq. (2.11a) that at high values of Q_1, $\Theta = 0$. This corresponds to capture of all emitted electrons by acceptors and absence of the reverse diffusion current.

At low Q_1 values ($Q_1 x_0 \ll 1$), when the phase shift is more essential, we obtain from Eq. (2.11a)

$$\tan \Theta = \frac{\Omega/k_s + Q_2}{k_A c_A/k_s + Q_1} \tag{2.30}$$

It should be noted that the phase shift Θ, being independent of x_0, is solely determined by the relative values of $\mathscr{D}_e$, Ω, and rates of processes involving solvated electrons. The character of the frequency dependence of $\tan \Theta$ allows the rate of capture of solvated electrons by the metal to be evaluated as well as the rate of electron capture by acceptors. In the limit $k_s \to \infty$ we obtain $\tan \Theta = Q_2/Q_1$. The latter relation is similar to the corresponding expression for the phase shift between the current and potential under conditions of alternating current passing through an electrochemical cell in which a chemical bulk reaction occurs (Fetter, 1961).

If the products [eA] participate in an electrode reaction, the situation becomes more complex. Let the acceptors be hydrogen ions, for example. The [eA] product in this case is then the hydrogen atom H according to

$$e_{aq}^- + H_3O^+ \to H + H_2O \tag{2.A}$$

which can take part in subsequent electrode reactions:

$$H + H_3O^+ + e^-(M) \xrightarrow{\vec{k}_1, \vec{k}_2} H_2 + H_2O \quad \text{(electrochemical desorption)} \tag{2.B}$$

$$H + H_2O \xrightarrow{\overleftarrow{k}_1, \overleftarrow{k}_2} H_3O^- + e^-(M) \quad \text{(ionization)} \tag{2.C}$$

as well as participate in recombination at the electrode surface:

$$H_{ad} + H_{ad} \xrightarrow{k_r} H_2 \tag{2.D}$$

where $e^-(M)$ designates an electron accepted from or donated to the electrode. Reactions similar to (2.B) and (2.C) can, in general, proceed with dissolved or adsorbed hydrogen. The rate constants of the two paths are $\vec{k}_1, \overleftarrow{k}_1$ and $\vec{k}_2, \overleftarrow{k}_2$, respectively. The discussion in this chapter will be limited to the second case: reactions of adsorbed [eA] products.

In principle, the recombination of hydrogen atoms can also proceed in the bulk, $H + H \rightarrow H_2$; however, owing to the low bulk concentration of atomic hydrogen under conditions of photoemission experiments, bulk recombination is very slow and can be neglected.

Reactions (2.B) and (2.C) involve additional electron transfers. The product of electronic charge e and diffusion flux of [eA] at the electrode surface coincides with the electric current I_{eA} in Eq. (2.1) only in the absence of surface recombination. In the converse case

$$\mathscr{D}'(dc'/dx)|_{x=0} = \vec{k}_2\theta + k_r\theta^2$$

where $\vec{k}_2$ is the rate constant of electrochemical desorption [reaction (2.B)], k_r is the rate constant of recombination [reaction (2.D)], and θ is the coverage with adsorbed hydrogen. Only the first term contributes to the photocurrent, so that $I_{eA} = e\vec{k}_2\theta$.

The bulk concentration of [eA] under conditions of modulated illumination is obtained from Eq. (2.3a) as

$$\frac{d^2c}{dx^2} - Q'_\Omega c'(x) + \frac{1}{\mathscr{D}'} k_A c_A c_e = 0$$

where

$$Q'_\Omega \equiv (i\Omega/\mathscr{D}')^{1/2} = (\Omega/2\mathscr{D}')(1 + i)$$

If recombination of [eA] is negligible, the complex amplitude is given (Rotenberg and Gurevich, 1968) by

$$I_{eA} = k_A c_A [\mathscr{D}_e(1 + Q'_\Omega \mathscr{D}'/k_{eA})(Q_\Omega^2 - Q'^2_\Omega)]^{-1}$$

$$\times e \int_0^\infty \left[e^{-Q'_\Omega x} - \frac{1 + Q'_\Omega \mathscr{D}_e/k_s}{1 + Q_\Omega \mathscr{D}_e/k_s} e^{-Q_\Omega x} \right] \Phi(x) \, dx \qquad (2.31)$$

Under stationary conditions, when $\Omega = 0$, we have $Q'_\Omega = 0$, so that Eq. (2.31) reduces to $I_{eA} = I - I_e$.

At high acceptor concentrations ($Q_1 x_0 \gg 1$), when one of the reactions of type (2.B) or (2.C) proceeds at the electrode, Eqs. (2.1) and (2.31) yield

$$j = I\left(1 \pm \frac{1 - Q'_\Omega x_0}{1 + Q'_\Omega \mathscr{D}'/k_{eA}}\right) \qquad (2.32)$$

The "+" sign refers to reduction, and "−" to oxidation of the products [eA] at the electrode. In particular, when oxidation of [eA] is so fast that $Q'_\Omega \mathscr{D}'/k_{eA} \ll 1$, we obtain from Eq. (2.32) $j = Q'_\Omega x_0 I$ and $\tan \Theta = 1$. Under conditions of slow oxidation $\tan \Theta = k_{eA}/(2\mathscr{D}'\Omega)^{1/2}$ [cf. Eq. (2.32)]. For fast and slow reduction, we obtain the expressions $\tan \Theta = x_0(\Omega/8\mathscr{D}')^{1/2}$ and $\tan \Theta = - k_{eA}/(2\mathscr{D}'\Omega)^{1/2}$ respectively. Thus, the magnitude and sign of the phase shift depend on the rates and nature of the electrode reactions involving [eA].

Similar analyses can be carried out with respect to other special cases which follow from Eq. (2.31) for $Q_1 x_0 \ll 1$. These were made by Rotenberg and Gurevich (1968).

Alternating photocurrents of another origin, namely those arising under conditions of constant illumination but with the electrode potential modulated in an oscillating manner, were also discussed by Rotenberg *et al.* (1968b). The photoimpedance of electrochemical systems $\hat{Z}_{ph}$ is determined by $\varphi^{(0)} = \hat{Z}_{ph} j^{(0)}$, where $\varphi^{(0)}$ and $j^{(0)}$ are the amplitudes of oscillating potential and photocurrent, respectively. The quantity $\hat{Z}_{ph}$ can have both capacitive and inductive components.

The Case of Intermittent Light

Golub (1969) calculated the photocurrent under conditions of intermittent illumination of an electrode assuming that the [eA] products undergo no discharge at the electrode which captures solvated electrons at an infinite rate. Also, assuming, as previously, that the decrease of bulk concentration of the acceptor during the experiment can be neglected, the problem reduces to the solution of Eq. (2.3) with the boundary conditions $c_e(0, t) = c_e(\infty, t) = 0$ and the initial condition $c_e(x, 0) = 0$. Upon integration, the photocurrent $j = I - I_e$ is obtained in the form

$$j = e \int_0^t \int_0^\infty \Phi(x, t) \left[1 - \frac{x \exp\{-k_A c_A(t - t') - x^2/4\mathscr{D}_e(t - t')\}}{[4\pi\mathscr{D}(t - t')]^{1/2}}\right] dx \, dt'$$

(2.33)

For periods of time considerably exceeding the solvation time (10^{-12} sec) we have approximately $\Phi(x, t) = \Phi(x)g(t)$, where $g(t)$ is a specified function describing the time dependence of light intensity. In particular, if $g(t) = \theta(t)$, where $\theta(t)$ is a step function (cf. p. 22), then

$$j(t) = e \int_0^\infty \Phi(x) \left[1 - \frac{1}{2}\left\{\exp(-x\sqrt{k_A c_A/\mathscr{D}_e}) \, \mathrm{erfc}\left(\frac{x}{\sqrt{4\mathscr{D}_e t}} - \sqrt{k_A c_A t}\right)\right.\right.$$
$$\left.\left. + \exp(x\sqrt{k_A c_A/\mathscr{D}_e}) \, \mathrm{erfc}\left(\frac{x}{\sqrt{4\mathscr{D}_e t}} + \sqrt{k_A c_A t}\right)\right\}\right] dx$$

(2.34)

Equation (2.34), together with measurements of $j(t)$ can, in principle, supply information concerning the shape of the source function $\Phi(x)$.

Barker and Gardner (1973b) considered stepwise pulsing of the illumination for conditions where a sufficiently fast reaction of [eA] takes place at the electrode. The acceptor concentration is sufficiently small, so that $x_0(k_A c_A/\mathcal{D}_e)^{1/2} \ll 1$; then the source of solvated electrons is effectively located in the $x = 0$ plane. Accordingly, Eqs. (2.3) and (2.3a) are solved with the following boundary and initial conditions:

$$
\begin{aligned}
t < 0: &\quad c_e(x) = c'(x) = 0 \\
t > 0: &\quad c_e|_{x=0} \equiv c_e^0, \qquad c_e|_{x=\infty} = 0 \\
&\quad c'|_{x=0} = c'_{x=\infty} = 0
\end{aligned}
\tag{2.35}
$$

The quantity c_e^0, which characterizes the source, depends on I and, in general, on the rate of electron capture by the surface. It can be found by solution of the corresponding equation for $x \simeq x_0$; c_e^0 appears in this problem as an extrinsic parameter.

Using Eqs. (2.3) and (2.3a), the nonstationary diffusion current of [eA] to the electrode and the nonstationary current of solvated electrons from the electrode can be calculated. The experimentally observed photocurrent is a combination of those two quantities. Assuming $\mathcal{D}' = \mathcal{D}_e$, we obtain

$$
\begin{aligned}
j(t) = c_e^0 e \mathcal{D}_e^{1/2} \Big\{ & (\pi t)^{-1/2} + (k_A c_A)^{1/2}(1 - \nu_{eA})\, \mathrm{erf}(k_A c_A t)^{1/2} \\
& - \nu_{eA}\left(\frac{\mathcal{D}_e}{\mathcal{D}_e - \mathcal{D}'}\, k_A c_A\right)^{1/2} \exp\left(\frac{\mathcal{D}' k_A c_A t}{\mathcal{D}_e - \mathcal{D}'}\right) \\
& \times \left[\mathrm{erf}\left(\frac{\mathcal{D}' k_A c_A t}{\mathcal{D}_e - \mathcal{D}'}\right)^{1/2} - \mathrm{erf}\left(\frac{\mathcal{D}_e k_A c_A t}{\mathcal{D}_e - \mathcal{D}'}\right)^{1/2} \right] \Big\}
\end{aligned}
\tag{2.36}
$$

Equation (2.36) becomes simplified considerably for time periods corresponding to $k_A c_A t \gg 1$. We obtain then from Eq. (2.36)

$$
j(t) = c_e^0 e(\mathcal{D}_e k_A c_A)^{1/2} \left\{ (1 - \nu_{eA}) - \nu_{eA}\sqrt{\frac{\mathcal{D}_e}{\mathcal{D}'}}\,(\pi k_A c_A t)^{1/2} \right\}
\tag{2.37}
$$

It follows from Eq. (2.37) that from the measured time dependence of the photocurrent $j(t)$, the ratio $\mathcal{D}_e/\mathcal{D}'$ can be obtained from $j \simeq t^{-1/2}$ plots if $k_A c_A$ and ν_{eA} are known. Furthermore, if $\mathcal{D}_e$ (or $\mathcal{D}'$) is known, $\mathcal{D}'$ (or $\mathcal{D}_e$) can be calculated.

The same paper describes a calculation of the photocurrent $j(t)$ for a "stepwise" switching on of light under conditions corresponding to $\Phi(x) = I/e \cdot \delta(x - x_0)$, assuming solvated electrons to be captured sufficiently fast for $c_e(0, t) = 0$ to apply. The expressions obtained in this case are rather tedious and will not be presented here. For $t \to \infty$, the photocurrent is given by $j = I[1 - \exp(-x_0 Q)]$; the latter equation corresponds to Eq. (2.11) when $\Phi(x)$ is replaced by the expression presented above, assuming $k_s \to \infty$ and $\nu = 1$.

Equivalent Electrical Circuits

It was already mentioned that quantitative discussion of electron transfer processes in electrochemistry can be approached not only analytically, but also by calculations involving equivalent electrical circuits. The latter approach was used by Barker *et al.* (1966) for calculation of photodiffusion currents.

As an illustration, the simplest circuit corresponding to the model of formation of hydrated electrons in a single plane at the distance x_0 from the electrode is shown in Fig. 2.1. Electron capture is assumed to occur only in the region $x > x_0$. The formation of the solvated electron cloud can be represented as a current generator I located at $x = x_0$. Diffusion of electrons is equivalent to the motion of charge along a half-infinite RC line with resistance (per unit length) r. Each subsequent resistance r_e is shunted by a capacitance (not shown in Fig. 2.1). The homogeneous capture of electrons by acceptors is represented by shunting resistances r_A uniformly distributed along the line for $x > x_0$. The measured photocurrent equals the difference between the current I and the current in the $x = 0$ plane. The resistance r in the $x = 0$ plane differs from zero if the electron capture by the surface proceeds at a finite rate. Using similar schemes, Barker *et al.* (1966) derived a number of expressions for the measured photocurrent.

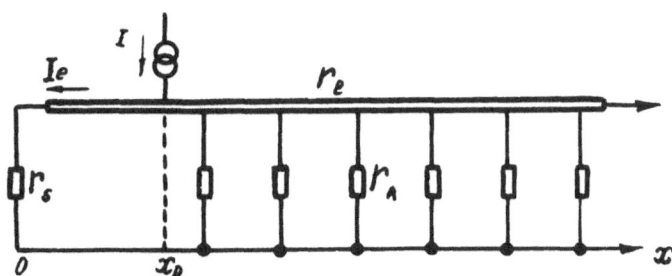

Fig. 2.1. Equivalent electrical circuit for a photodiffusion process (Barker *et al.*, 1966).

Experimental Techniques in Photoemission Studies

3.1. General Discussion of Photocurrent Measurements

The photoemission process is usually characterized by its quantum yield Y referred to the given frequency of the perturbing light. It represents the ratio of the number of emitted electrons to the number of photons incident (and sometimes absorbed) on the electrode. Measurements of the quantum yield require a precise determination of the power of the light flux incident (or absorbed) per unit area of electrode surface. The quantum yield of single-photon emission in vacuum excited by quanta with energies of the order of a few electron volts usually does not exceed $10^{-2}-10^{-3}$ electron/photon.

The solution of the majority of problems considered in the text which follows requires only comparative measurements of the photoemission intensity under varying experimental conditions (in particular, at various electrode potentials). In other words, there is no need to establish the absolute magnitude of the quantum yield. It is sufficient to measure the photoemission current, proportional to Y, under conditions of constant illumination during the entire cycle of experiments.

The photoemission experiments to be described further consist in measurements of photocurrent (sometimes photopotential) in an electrochemical cell with one illuminated electrode as a function of potential, frequency and intensity of incident light, acceptor concentration, etc.

Methods of measurement, as well as the choice of the photoactive electrochemical system (i.e., electrode material, solution composition, etc.), depend on the aims of experiments. If possible, side photoprocesses, enumerated in Section 0.1, should be avoided.

The apparatus for photoemission studies can be arbitrarily divided (depending on the mode of illumination and, thus, on the construction of the measuring circuit) into two types: (a) that operating under conditions of a constant, time-independent signal, and (b) that operating under conditions of an alternating signal. In the first case, the electrode is illuminated with light of constant intensity and dc measuring apparatus is used. In the second case, light intensity is varied with time in a prescribed way (e.g., sinusoidal, square wave, pulse signal); then the recording of the effects requires the use of ac amplifiers.

Modulation of the exciting light has the following advantages: (1) Alternating current amplifiers are in many respects more convenient to use than dc systems and they allow the background (dark) current to be easily distinguished from the measured photosignal. (2) Processes can be studied which result in a phase shift between the photocurrent and incident light intensity [e.g., due to diffusion of reactants or intermediates, inhibited reverse current of solvated electrons, etc. (cf. Chapters 2 and 5)]. The frequency modulation necessary for use of ac amplifiers does not usually exceed 100 Hz — it is limited at the upper end by the band-pass of the potentiostat used to maintain the electrode at a constant potential. Studies of relaxation processes, however, require much higher modulation frequencies. In fact, the characteristic time of processes involving bulk reactions with acceptors is equal to $1/k_A c_A$ (where k_A is the rate constant of solvated electron capture, and c_A is the acceptor concentration). For $k_A = 10^9$ liters/mole-sec and $c_A = 10^{-5}$ mole/liter, the characteristic time is close to 10^{-4} sec. The return to the electrode of solvated electrons, which had not been captured by acceptors, is even faster, of the order of $\chi_0^2/\mathscr{D}_e$, where χ_0 is the mean solvation length and $\mathscr{D}_e$ is the diffusion coefficient of the solvated electron. For $\chi_0 \simeq 10^{-7}$ to 10^{-6} cm and $\mathscr{D}_e = 5 \cdot 10^{-5}$ cm²/sec, the return time does not exceed 10^{-8} sec. Thus, the modulation frequency used in studies of relaxation processes should be of the order 10^4 to 10^8 Hz.

One of the following methods of light modulation is usually employed: a mechanical method, using a rotating disc with holes placed in the path of the light beam; an electrical method consisting in prescribed variations of the current supplied to the lamp, or in use of pulsed power light sources; an electro-optical method consisting in directing the light beam through an electro-optical element whose transmission is regulated by an applied electric field [e.g., the Kerr cell (cf. Mustel' and Parygin (1970)]. Borbat *et al.* (1967) and Litvak (1966) described some additional methods of light modulation. Low-frequency modulation (10–1000 Hz) can be effected by the first method, which is, however, inapplicable in the case of fast processes, for which short-period flash lamps or lasers are required. Electro-optical devices with fast response are not yet widely used in photochemistry, owing mainly to the difficulty of attaining the necessary transmission in the ultraviolet region.

Comparison of the theory with experiment (Chapter 1) requires the photoemission current density to be known. It can be measured directly at constant electrode potential, independent of light intensity variations. This condition is easy to satisfy under stationary conditions of illumination or with low-frequency light modulation, by using potentiostats of various types.

However, for high-frequency and pulse measurements, when the electrode impedance is very low (owing to the reactive conductivity component), potentiostats should have a very low input resistance; construction of such potentiostats involves serious difficulties. Therefore, measurements are usually carried out galvanostatically instead of potentiostatically, i.e., at a constant current in the electrochemical cell. (The coulostatic method is a special case of the galvanostatic method, when the cell current equals zero.) The photopotential φ_{ph} is then the measured quantity.

In order to clarify the above description it is useful to consider Fig. 3.1. It represents an equivalent circuit of a cell with a photosensitive electrode (the impedance of the counter-electrode is negligibly small). The photoelectrode itself is represented by a parallel connection of the photocurrent generator I, a capacity C, and a reaction resistance R. Transition from galvanostatic to potentiostatic conditions is determined by the ratio of the photoelectrode impedance $\hat{Z}$ (at the given frequency) and the external resistance which includes the resistance of the solution, R_{el}, the polarization circuit, R_n, and of the measuring device, R_0. Under stationary conditions $\hat{Z} = R$, and potentiostatic conditions obtain if $R \gg R_{el} + R_n + R_0$. In the reverse case the current remains constant in the external circuit, but the electrode capacity C becomes charged, and illumination results in the appearance of the photopotential φ_{ph}. The measured φ_{ph} value allows the "emitted charge" $Q = \int j(t)\, dt$ (i.e., the change of the electrode charge during illumination) to be calculated by integration of the photocurrent j. It is particularly easy in the case of an ideally polarizable electrode; it is virtually equivalent in this case to maintaining the condition $RC \gg t$ (where t is the flash time); cf. the equivalent circuit in Fig. 3.1.

The capacity of the double layer is usually measured together with the photopotential using the same apparatus. The system is perturbed, however, not by light but by a signal from a suitable voltage generator (see below). Difficulties arise if the capacity is frequency (i.e., time) dependent, as often

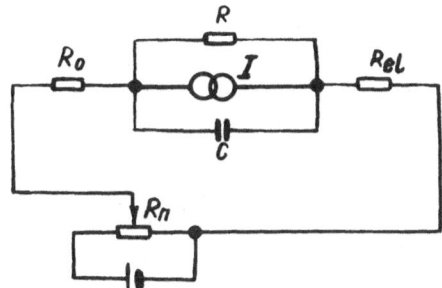

Fig. 3.1. Equivalent electrical circuit of an electrochemical cell with a photosensitive electrode and measuring circuit.

happens in the case of solid electrodes, adsorbed organic substances, and in other situations. The error connected with the frequency dispersion of capacity can be decreased by making frequency spectra of both probe signals — light for photopotential measurement and electrical for capacity measurement — coincide. This is not always easy to achieve experimentally.

If the potential is kept constant by means of a potentiostat, the equivalent circuit remains the same (Fig. 3.1), but instead of the full external resistance ($R_{ex} = R_{el} + R_n + R_0$) it is replaced by the solution resistance between the test (illuminated) electrode and the tip of the Luggin capillary connected with the reference electrode.

Among various methods for electrode illumination and measurement of the photosignal, stationary illumination supplies the simplest way for a precise photocurrent measurement. Under favorable conditions, the total error connected with all experimental factors amounts to 2–3%. In the case of a large but time-independent background ("dark") current, partial compensation of the latter can decrease the errors. This cannot be done, however, in the case of time-dependent "dark" currents.

The accuracy of pulse measurements is relatively poor (mean error is usually about 10%). This is connected with the accuracy of the direct reading of the signal magnitude on the oscilloscope screen and with errors of capacitive measurements. Also, the operation of flash lamps is usually accompanied by electric discharge (in the lamp power supply circuit), resulting in significant electrical noise.

Periodic intensity modulation is the most universal illumination method. It combines high accuracy with the possibility of carrying out measurements in the presence of relatively high background currents. The method is particularly important for solid electrodes which usually have a rather narrow region of ideal polarizability.

3.2. Measuring Apparatus

One of the first types of apparatus used for modulated cell illumination, based on square-wave polarography, was described by Barker *et al.* (1966). Light intensity of a low- or high-pressure mercury lamp was varied by square-wave modulation of the lamp feed current with a half-cycle of 45 msec. A square-wave polarograph was used for cell polarization and photocurrent measurements. The latter were started 25 msec after switching on of the first half-cycle. The photocurrent density was 10^{-7} to 10^{-5} A-cm^{-2}. (Photocurrents measured under conditions of constant light intensity are usually of the same order of magnitude.)

An apparatus generating a triangular light signal was described by

De Levie and Husovsky (1969) and by De Levie and Kreuser (1969). The dropping mercury electrode was used as the photocathode. This ensures the best reproducibility of the electrode surface. Potential is varied slowly and linearly in time, each new photocurrent measurement being carried out at a new drop (at a given moment in its lifetime). A simplified block diagram of the apparatus is shown in Fig. 3.2a. The cell current varies owing to external perturbations either by modulation of the light or (for impedance measurements) by a sinusoidal signal from an ac generator. Current is measured and recorded using a narrow-band amplifier tuned to the modulation frequency, and a lock-in detector with a pen recorder.

The optical diagram of the apparatus is shown in Fig. 3.2b. Two disc interruptors are placed in the path of the light beam. One of them modulates the intensity of incident light with a frequency of 16 Hz. This low measuring frequency allows the use of potentiostats with narrow operating band and, consequently, with highly accurate constancy of potential. This interruptor, together with the auxiliary light source, provides a reference signal (for the lock-in detector) of the same frequency as the measuring one. The second interruptor, with 0.1 Hz frequency, serves to synchronize the recorder with the mechanism, forcing the drop to detach. The recording pen touches the tape at a prescribed moment after detachment of the former drop. Thus, each new drop provides a single point on the photocurrent vs. potential curve. The potentiostatic circuit, as well as the phase-sensitive device for measuring the active and reactive components of the total admittance of the interface and of the compensator of the ohmic solution resistance, are described in the papers cited.

Means and Mark (1972) measured the photocurrent using a phase-sensitive detector, finely tuned to the modulation of the light frequency. In this way the noise was considerably decreased [to 10^{-11} A for the background ("dark") current of 10^{-5} A].

The light modulation is unnecessary if flash lamps or lasers are employed. The cell signal (usually photopotential) is measured then by means of wide-band amplifiers.

The coulostatic arrangement, described by Delahay and Srinivasan (1966), works in the following way. A mercury pool electrode is polarized to the desired potential (at which no electrode reactions proceed with any significant rates). Then the polarization circuit is interrupted and the electrode is illuminated immediately for about 1 msec by a flash lamp. Potential decay is recorded by an oscilloscope with a differential input, connected with another, nonilluminated electrode, identical with the test one. The differential technique removes the error due to low background currents which pass during the flash time and the time of potential reading, and disturb the coulostatic mode of operation.

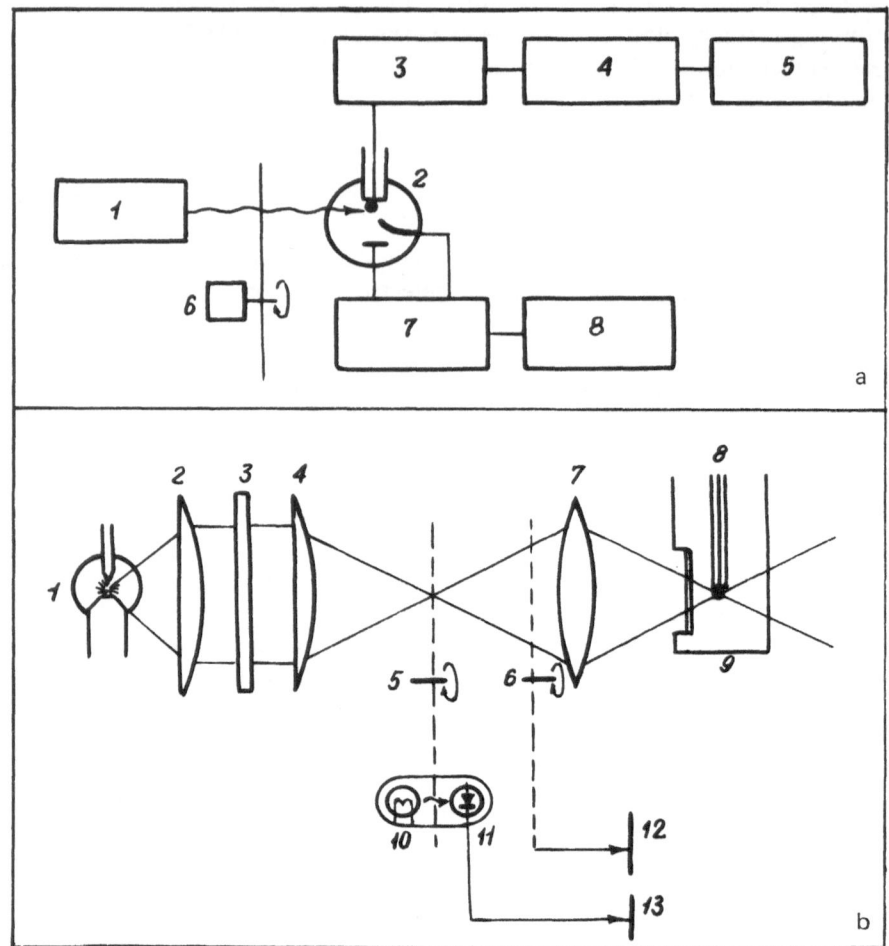

Fig. 3.2. Apparatus with modulated electrode illumination (De Levie and Kreuser, 1969). (a) Block diagram. 1, light source; 2, electrochemical cell; 3, current amplifier; 4, lock-in detector; 5, *X-Y* recorder; 6, light interrupter; 7, potentiostat; 8, ac generator. (b) Optical diagram. 1, mercury lamp; 2, 4, 7, lenses; 3, light filter; 5, 6, light interruptors; 8, dropping mercury electrode; 9, electrochemical cell; 10, auxiliary lamp; 11, photodiode; 12, input of the recorder synchronization unit; 13, input of the lock-in detector reference signal.

The measuring procedure can be conveniently illustrated in terms of the schematic in Fig. 3.3a. The potential of the test electrode e_1 (illuminated test electrode) and of the reference electrode e_2 is imposed using a potentiometer P_1 and a counterelectrode e_3 with the switch S in position 1. The circuit

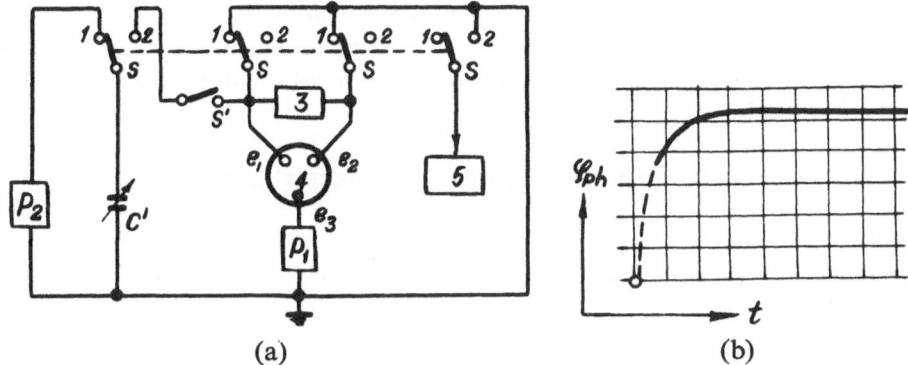

(a)   (b)

Fig. 3.3.   Arrangement for pulse-illuminated electrode (Delahay and Srinivasan, 1966). (a) Block diagram. e_1, photoelectrode; e_2, reference electrode; e_3, auxiliary electrode; S, S′, switches; P_1, P_2, potentiometers; C′, condenser; 3, oscillograph; 4, electrochemical cell; 5, triggering unit for the flash lamp. (b) Photopotential oscillogram. Mercury electrode, 0.1 N HCl + 0.4 N NaCl solution. Horizontal scale unit, 0.5 msec; vertical scale unit, 10 mV.

is interrupted by switching S to position 2. The flash lamp and the oscillograph sweep are simultaneously triggered. (A characteristic oscillogram is shown in Fig. 3.3b.) The plateau shown in the oscillogram after the flash has ended proves that the necessary condition (see above) $RC \gg t$ (where t is the flash time) was experimentally satisfied. The latter condition prevents the faradaic "escape" of charge from the electrode during the measurement. The time independence of photopotential shows that the electrode capacity is constant within the frequency range studied.

The magnitude of the emitted charge is measured as follows. By switching S in position 1, the potential between electrodes e_1 and e_2 is set at the same value as that before the light flash. When S is switched to position 2, the condenser with capacity C′ discharges onto the electrode e_1. It should be mentioned that the initial potential difference on the discharging condenser C′ should exceed considerably the rational potential (referred to the zero-charge potential) of the test electrode. Otherwise, condenser C′ does not fully discharge into the double-layer capacity. The charge of condenser C′ is chosen (by means of the potentiometer P_2) in such a way that the potential change of electrode e_1 observed upon charging is equal to the photopotential. Thus, the differential electrode capacity need not be explicitly determined. It is important, however, that the electrode capacity be the same in the dark and under illumination (as is usually the case; cf. De Levie and Kreuser, 1969; Berg and Reissmann, 1970).

A pulse apparatus for photoelectrochemical measurements was also described by Korshunov *et al.* (1971). No potentiostat was used and the electric circuit was not interrupted during illumination. The photopotential φ_{ph} was measured on a standard resistance R_0 connected in series with the cell. The photocurrent is given by

$$\varphi_{\text{ph}}(t) = \exp(-t/CR_{\text{ext}}) \int_0^t C^{-1} j(t') \exp(t'/CR_{\text{ext}}) \, dt'$$

where t is time, $R_{\text{ext}} = R_{\text{el}} + R_n + R_0$, C is the double-layer capacity, and t' is the integration variable. For pulse times much less than CR_{ext}, we obtain

$$\varphi_{\text{ph}}(t) = \int_0^t C^{-1} j(t') \, dt'$$

i.e., the photopotential is proportional to the charge emitted by the electrode.

The flash times for lasers used as the photopower source are very short, of the order of 10^{-8} sec. This poses certain difficulties connected with the sufficient accuracy of measurements of energy radiated during the flash and with synchronization of the measuring circuit with the flash. Figure 3.4 illustrates a block diagram of such a system (Benderskii *et al.*, 1974). The light from a ruby laser 1 passes through a light filter 2, polarizer 3, and focusing lens 4, and illuminates the electrode in cell 5. Semitransparent mirrors 6 deflect a part of the beams onto the diode 7, which triggers the double-beam oscillograph 11, and onto the photocell 8, for measurements of the peak intensity and energy of the flash (using calorimeter 9). The photocell signal is fed through the integrator 12 into the oscillograph input. The signal of the cell photopotential obtained from the resistance R_0 (about 10^3 Ω in series with the cell) and amplified by an amplifier 10 is fed to the

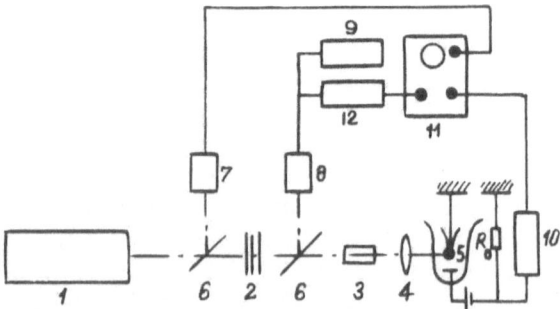

Fig. 3.4. Block diagram of an arrangement with laser-illuminated electrode (Benderskii *et al.*, 1974). 1, ruby laser; 2, light filters; 3, polarizer (Glan prism); 4, lens; 5, electrochemical cell; 6, semitransparent mirror; 7, photodiode; 8, photocell; 9, calorimeter; 10, amplifier; 11, double-beam oscillograph; 12, current integrator.

second oscillograph input. Simultaneous control of the characteristics of the light flash and photopotential allows satisfactory accuracy of measurements to be attained, regardless of the reproducibility of the peak intensity and flash energy.

3.3. Electrode Illumination

Photoemission experiments require, as a rule, monochromatic light. The spectral range is limited on the high-frequency side by the effects of light absorption by solution and cell materials, and on the low frequency side by the magnitude of the electronic work function in solution. The electronic work function of metals in vacuum, which is of interest from the photoelectrochemical point of view, is usually 4–5 eV. The work function in solutions is lower (in aqueous solutions, e.g., by 1.25 eV) owing to the electron interactions with the medium (cf. Section 4.4). Additional decrease of the work function is caused by the cathodic polarization of the electrode. Therefore, the working range of quantum energies is about 2.5–5 eV and, thus, includes the visible and near ultraviolet parts of the spectrum. Monochromatic light is obtained from lamps with line or continuous emission spectra by means of monochromators, or, more often, using light filters.

Lasers used in photoelectrochemical studies are usually of the solid type (e.g., the ruby laser) since the power of ultraviolet gas lasers is still too low. The energy of the quantum of a ruby laser corresponding to the basic frequency is too low to excite single-photon emission, and a special optical device is used to include the second or third harmonics.

The most commonly used lamps in photoemission, and generally in photochemical experiments, are low, high, or ultrahigh pressure mercury lamps with a quartz body (cf., e.g., Rokhlin, 1966). Low-pressure lamps have a line spectrum and are especially well suited for generation of short-wave light (e.g., 2540 Å). The line spectrum of high or ultrahigh pressure lamps is superimposed on a continuous background, the intensity of which increases with increasing mercury vapor pressure. These lamps have, as a rule, small dimensions (several millimeters) which facilitate light focusing. Xenon arc lamps have a continuous spectrum. Relaxation measurements are carried out using flash lamps with flash energies of a few hundreds of joules and flash times from 10^{-3} to 10^{-6} sec (Marshak, 1963; Rokhlin, 1966).

The methods for obtaining monochromatic light will now be discussed. The use of monochromators is somewhat restricted since the quantum yield of the photoelectron emission at the metal/solution interface is usually 10^{-4} to 10^{-3} electron/photon. At the same time, the electrochemical photocurrents should be considerably higher than vacuum photoeffects, owing to the existence of background faradaic currents (usually of the order

10^{-7} to 10^{-6} A-cm^{-2}). Photocurrents of this magnitude require strong illumination. Standard monochromators, however, have, as a rule, insufficient transmission and therefore cannot be employed as widely in photoelectrochemical experiments as in vacuum photoemission studies.

In practice, the main device used for obtaining "monochromatic" light in photoelectrochemical studies are interference light filters, which allow high-intensity illumination and a sufficiently narrow spectral range to be attained simultaneously. Their maximum transmission is 30–50%, and the bandwidth at the half-maximum height is 100–150 Å. They are usually employed together with glass light filters which cut off the "unnecessary" part of the spectrum; a detailed description of such combined arrangements can be found in the works of Koller (1965) and Parker (1968). The heating of the electrode and solution is prevented by cutting off the long-wave part of the spectrum with water filters.

The light is focused using quartz optics (up to 365 nm uviol glass can be used). Methods of obtaining polarized light have been described by Shurkliff (1962).

Light intensity can be measured using photomultipliers and vacuum photocells sensitive to the desired spectral range (cf., e.g., Korndorf et al., 1967). They are usually employed in comparative measurements. In certain cases, e.g., determination of the photoemission quantum yield, studies of the spectroscopic and other characteristics of the process require light detectors calibrated over a sufficiently wide spectral range. In spite of their low sensitivity, radiothermoelements are most convenient for these purposes. Sometimes photomultipliers with windows covered with a luminophor layer are also used [Zaidel' and Shreider (1967)].

3.4. Electrochemical Cells

Early photoelectrochemical studies were usually carried out using mercury electrodes. The mirrorlike smoothness and reproducibility of its surface makes mercury a "model" electrode in photoelectrochemistry as in other branches of electrochemistry. Accumulation of impurities is prevented by using the dropping mercury electrode, which has been described in detail in the polarographic literature (cf., e.g., Heyrovsky and Kuta, 1962). Also a hanging (renewable) drop electrode [a convenient arrangement as described by Gokhshtein and Gokhshtein (1962)] or mercury in the form of a flowing pool can be used. The latter type is used if the experiment requires a flat electrode surface, as is the case, for example, in studies of the effect of polarization of light on photoemission phenomena.

Cells are constructed of quartz or are fitted with a quartz window for illumination of the electrode.

Photoemission studies do not require, in principle, any special pre-treatment of solid electrodes except for the careful removal of oxide layers which roughen the surface and may give rise to photocurrents due to photoconductive effects, photosensitization of electrode reactions, etc.

The solution composition is determined by the purposes of the experiment. It must, however, be taken into account that absorption of light by solution components decreases illumination of the electrode and may complicate the conditions owing to the appearance of photolysis products in the bulk solution. Finally, the choice of solution should allow for a sufficiently wide range of potentials corresponding to a good polarizability of the electrode. Requirements with respect to the purity of materials used and experimental methods do not differ from those set by other branches of theoretical electrochemistry.

3.5. The Choice of Acceptors for Solvated Electrons

It was already mentioned that stationary photocurrents due to electron emission can be observed only in the presence of acceptors of solvated electrons, although (cf. Section 0.2) solvated electrons can also react with solvent molecules as exemplified by the reaction

$$e_{aq}^- + H_2O \rightarrow H^\cdot + OH^-$$

The rate of this reaction is, however, so low that it can give rise only to extremely small (rest) currents.

In spite of the large number of substances which effectively react with solvated electrons, the number of acceptors suitable for photoemission measurements is relatively small. An "ideal" acceptor should satisfy several requirements: (a) it should have a sufficiently high rate constant of its reaction with solvated electrons; (b) it should have sufficient solubility in the solution studied; (c) it should not participate in any electrode reactions in the potential range studied; (d) it should absorb no light in the light-frequency range studied; and (e) the capture product [eA] should not become oxidized at the electrode, otherwise the photocurrent will be close to zero. Obviously, this does not pertain to studies of acceptor oxidation or reduction *per se* (cf. Chapters 7 and 8).

Often it is desirable that the scavenger should not be strongly (specifically) adsorbed at the electrode–solution interface in the investigated potential range, otherwise complications and uncertainties caused either by the capture of hydrated electrons by adsorbed scavenger molecules (or ions) or by light-induced electron transfer between the electrode and adsorbed entities (due to mechanisms of the types suggested by Heyrovsky and Berg) may arise.

Thus, the choice of the most suitable acceptor is determined by the actual physicochemical properties of the system studied (electrode material, solvent, solution composition) and by the aims of the investigation. The effectiveness of acceptors can be evaluated on the basis of tabulated rate constants of acceptor interactions with solvated electrons in aqueous and nonaqueous solutions, which have been previously reported in a number of monographs and reviews [e.g., Anbar and Neta (1967); Hart and Anbar (1970); Pikaev (1969)].

The acceptors most widely used in aqueous solutions are hydrogen ions, H_3O^+, nitrate ions, NO_3^-, nitrite ions, NO_2^-, and nitrous oxide, N_2O, although none of these satisfies the requirements of an ideal acceptor. Hydrogen, nitrate, and nitrite ions (as well as several organic molecules) have high rate constants of interaction with electrons and high solubility in water. However, their capture products can become oxidized at the electrode, so that they therefore can be used only within a rather narrow range of negative electrode potentials. (Reactions proceeding during photo-electron emission in acid solutions are discussed in Section 7.1, and those in nitrate solutions are examined in Section 8.1.) Moreover, concentrated nitrate and nitrite solutions absorb light to a considerable extent at wavelengths below 3000 Å.

The most convenient acceptor in aqueous solution is N_2O (the rate constant of the capture of hydrated electrons by N_2O is $k_A = 6 \cdot 10^9$ mole^{-1}-liter-sec^{-1}). N_2O does not undergo cathodic reduction at the mercury electrode down to -1.8 V and does not adsorb at its surface (De Levie and Kreuser, 1969). It can, however, adsorb on solid metals (lead, cadmium, indium, bismuth); see, for example, Eletskii *et al.* (1969).

The interaction of N_2O with hydrated electrons is usually assumed to occur through the following homogeneous and heterogeneous reactions:

$$N_2O + e_{aq}^- \rightarrow N_2O^- \qquad \text{(in the solution)}$$
$$N_2O^- + H_2O \rightarrow N_2 + OH^\cdot + OH^- \qquad \text{(in the solution)}$$
$$OH^\cdot + e^-(M) \rightarrow OH^- \qquad \text{(at the electrode)}$$

The final product of the homogeneous process, the $OH^\cdot$ radical, is reduced rapidly on mercury and other metals in the potential range normally applied in photoemission current measurements (-0.2 to -1.7 V). (The observation of Levin and Delahay, 1970, concerning formation of a product oxidizable at the electrode was not confirmed in a subsequent study by Barker and Bottura, 1973b.) The main disadvantage of N_2O is its limited solubility in water and aqueous solutions, which amounts to only 0.025 mole-liter^{-1} at room temperature and 1 atm pressure and is thus insufficient to enable all emitted electrons to be captured and attain the emission current.

Unless otherwise stated, further figures refer to photocurrents measured in arbitrary units, and to potentials measured with respect to the saturated calomel electrode and the wavelength of the incident light equal to 3650 Å.

Photoelectron Emission in Solutions: Its Discovery, Kinetics, and Energetics

4.1. Introductory Notes

At a certain stage of development of photoemission studies, the key question was how to prove the existence of photoemission in solutions, i.e., unambiguously distinguish this type of photoeffect from others enumerated in part in Section 0.1. The problem is less simple than it appears. The first direct experimental evidence of electron photoemission at the metal–vacuum interface was reported in the classic papers of Lenard (1900) and Thomson (1899) who showed that illuminated metals emit in vacuum the same particles as those previously found in cathode rays, i.e., electrons. Similar, direct evidence of electron photoemission in solution would, obviously, require a demonstration of the existence of solvated electrons in the vicinity of the electrode. However, it should be remembered that as opposed to vacuum where free electrons can exist for infinite periods of time, solvated electrons in electrolytes disappear relatively fast, being captured either by acceptors or by the electrode surface. Therefore, their stationary concentration at electrodes illuminated with usual intensities of light does not exceed 10^{10} to 10^{12} electrons per cubic centimeter. This is much below the sensitivity of modern physical methods for detection of hydrated electrons (absorption spectra, EPR, etc.).

Walker (1967a) tried to detect hydrated electrons (assumed to be formed in the process of thermal electron emission at a cathodic electrode polarization) in the vicinity of a cathode spectroscopically, i.e., through absorption by electrons of laser light tangentially scanning the electrode surface. As is clear now (from results of photoemission studies; cf. Section 10.3), this attempt was doomed to failure, since thermal emission is absent in solutions at the electrode potentials applied in Walker's experiments. The effects observed by him were proved later by Postl and Schindewolf (1971), Conway (1972), and Bewick et al. (1973) to be unconnected with formation of solvated electrons but due to the effect of potential on the surface reflectivity (so-called "electroreflection"; cf. Cardona, 1969). Thus, indirect methods are required to experimentally detect photoemission in solutions. The general criteria which, taken together, make possible the distinction between photoemission and other photovoltaic effects at the metal–electrolyte interface can be formulated as follows:

(a) A characteristic dependence of the photocurrent on potential and quantum energy.

(b) Independence of the photoemission threshold potential on the nature of the electrode metal (see below).

(c) The dependence of photocurrent on acceptor concentration in solution.

The first successful demonstration of the nature of the observed photocurrent as an emission process was made by Barker *et al.* (1966) in their study of the dependence of photocurrents on acceptor concentration. We shall, however, discuss first the effect of the electrode potential proceeding from criteria (a) and (b). The dependence of photocurrent on acceptor concentration will be discussed in the next section. It should be mentioned here that the basic photoemission concept requires a thorough experimental and theoretical study of a whole complex of spectral, voltammetric, and other characteristics of the photocurrent, since even such characteristic properties of the latter as the existence of the photoelectric threshold, rapid response, etc., may be interpreted separately in terms of almost all of the hypotheses discussed in Section 0.1 with respect to the possible origins of the photocurrent. It will be shown presently that the whole complex of accumulated experimental data has proved beyond doubt that electron photoemission is the predominant process in the photoeffect which occurs in the systems studied over a wide range of potentials and light frequencies. Simultaneously, the established laws of photoemission in solution have created a basis for the development of the present understanding of this phenomenon.

4.2. The Dependence of the Photoemission Current on the Electrode Potential and Quantum Energy. Experimental Verification of the 5/2 Power Law

Photoemission theories proposed up to 1967 predicted various types of dependence of photocurrent on potential and photon energy. The Fowler theory, which gives good agreement with experiment at the metal–vacuum interface, predicts a square law connecting the photocurrent with the energy of light quanta. When applied to the metal-electrolyte interface, together with expression (0.1) for the electronic work function in solution (cf. Sharma *et al.*, 1968), a similar (i.e., quadratic dependence) on potential would be expected. It follows, however, from the theory of Brodskii and Gurevich, developed for photoelectron emission into solutions (Section 1.4), that the dependence of the photoemission current I on potential and on the quantum energy $\hbar\omega$ is given by the 5/2 power law:

$$I = A(\hbar\omega - \hbar\omega_0(0) - e\varphi)^{5/2} \qquad (4.1)$$

where $\hbar\omega_0(0)$ is the work function for emission into solution at $\varphi = 0$; this quantity will be called in the text which follows simply $\hbar\omega_0$.

Theories which do not involve photoemission, viz. photodecomposition of the surface complex with charge transfer (Heyrovsky, 1966a) and "hot" electrode effects (Berg *et al.*, 1967), predict an exponential dependence of photocurrent on potential.

It must be mentioned here that the theory discussed in Chapter 1 embraces neither all conceivable mechanisms nor photoemission properties. Correspondingly, the 5/2 power law is not universal in the sense that it excludes all deviations. In fact, a considerable contribution to the photo-current from surface plasmons results in a more complex potential de-pendence of the photocurrent, as will be shown in Section 10.4. In some special cases which will not be discussed in detail, under conditions of photoexcitation of electrons in the metal bulk, deviations from the 5/2 power law can also, in principle, be expected. However, the experimentally observed dependence of the bulk-generated photocurrent on potential still obeys the 5/2 power law (Benderskii *et al.*, 1974). Therefore, in the con-tinuing text the problem of the bulk or surface origin of emitted electrons will not be considered for each separate case; it will, however, be discussed in detail in Section 4.5.

Already the first experiments on photoelectron emission into solution (Barker *et al.*, 1966; Delahay and Srinivasan, 1966) demonstrated the in-adequacy of Fowler's theory: The plots of $j^{1/2}$ vs. φ were not linear over the whole potential range. The same experimental data plotted by Gurevich *et al.* (1967) as $j^{0.4}$ vs. φ gave straight lines over a wide potential range.

The first experimental verification of the 5/2 power law was made by Korshunov *et al.* (1968a), who used pulsed illumination but did not take into account the potential dependence of the electrode capacity. A subsequent, more thorough, although still not strictly accurate, study of the potential dependence of photocurrent was carried out by Rotenberg and Pleskov (1968) and by De Levie and Kreuser (1969). The latter authors measured the photocurrent at a mercury electrode illuminated by modulated light (cf. Section 3.2) using N_2O, in a saturated solution, as the electron acceptor. Their results are illustrated in Fig. 4.1, in which $j^{1/2}$ (according to Fowler), $j^{0.4}$ (according to the 5/2 power law), and $\log j$ (according to Heyrovsky) are plotted against φ. It can be easily seen that only for the second case are

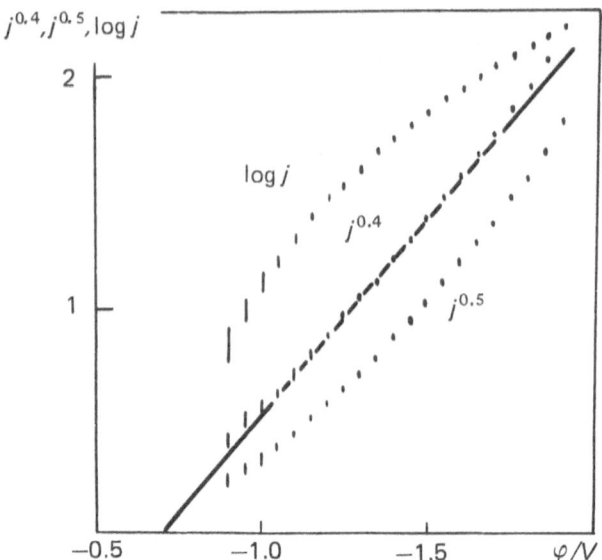

Fig. 4.1. Experimental photocurrent curves interpreted in terms of various theoretical laws (De Levie and Kreuser, 1969). Dropping mercury electrode in 0.1 M KCl solution saturated with N_2O. Wavelength 4360 Å.

the experimental results well approximated by a straight line. It can thus be concluded that the photocurrent observed is in fact of a photoemissive nature and can be well approximated by the 5/2 power law.

Before more accurate studies of the potential dependence of photo-current are described, it will be useful to consider requirements which should be satisfied by experimental conditions if a correct verification of the basic photoemission law is to be made. The first requirement is that the measured photocurrent j be equal, or at least close to, the emission current I (strictly speaking, νI) since it is not known *a priori* whether or not proportionality exists between the photocurrent measured at low acceptor concentrations and the emission current. Moreover, the stoichiometric coefficient ν must be constant over the entire potential range studied.

The first condition is satisfied if measurements are carried out at high acceptor concentrations. The most convenient acceptor is the hydrogen ion, whose concentration can be high enough (of the order of 1 mole-liter^{-1}) to ensure capture of virtually all emitted electrons. However, quantitative measurements can be carried out in strongly acidic solutions only at metals with high hydrogen overpotential. The potential range then accessible for quantitative measurements (to an accuracy of 3–5%) is approximately

-0.7 to -1.4 V. Outside this range, at more negative potentials, the photo-current is strongly affected by hydrogen evolution, while at more positive potentials it is affected, for example, by oxidation of lead or of atomic hydrogen on mercury. The latter reaction (discussed in detail in Chapter 7) causes the stoichiometric coefficient to continuously decrease at potentials more positive than -0.7 V from its constant value of 2 to zero.

The proof that experimental data obey the 5/2 power law, and not some other close relationship, e.g., Fowler's law, requires that two conditions be satisfied:

(1) Experimental data plotted as $j^{0.4}$ vs. φ should lie on a straight line, and the mean-square deviation from the linear dependence should obviously be less than that resulting from fitting the experimental data to other laws. Moreover, the mean-square deviation should be less than the mean-square experimental error.

(2) The threshold potential φ_0 determined by the intercept of plots of $j^{0.4}$ vs. φ on the potential axis and equal to $\varphi_0 = (\hbar\omega - \hbar\omega_0)/e$ should depend linearly on $\hbar\omega$, the slope of the $\hbar\omega$ vs. $e\varphi_0$ line being equal to unity.

As shown above, a verification of one or another theory can be made by statistical treatment of experimental data. The first attempt was made by Rotenberg *et al.* (1968a). Computer-calculated results confirmed that photo-emission into electrolytic solutions, as opposed to photoemission into vacuum, obeys the 5/2 power law rather than Fowler's law.

The most thorough investigation of this type was carried out by Benderskii *et al.* (1974). The voltammetric characteristics studied were represented by a step function

$$j = A(\varphi_k - \varphi_0)^n \qquad (4.2)$$

where φ_k is the measured potential (under experimental conditions $|\varphi_k - \varphi_0| \gg kT/e$ so that the "thermal tail" of the voltammetric charac-teristics in the vicinity of the threshold did not affect the results). The most probable value of the power n (as well as of the threshold potential φ_0 and the constant A) was determined by computer calculations using the least-squares method. The function

$$\mathscr{F}(\varphi_0, A, n) = \sum_k [j(\varphi_k) - j_k]^2 \qquad (4.3)$$

[where $j(\varphi_k)$ is the photocurrent calculated according to Eq. (4.2) and j_k is the experimentally measured photocurrent] was minimized using the configuration and Newton methods with respect to parameters A and φ_0 at various values of n within the range 2–3 fixed at 0.1 spacings. The lowest value $\{\min \mathscr{F}_n\}$ corresponds to the most probable value of the power n. The experimental dispersion of the photocurrent measurements, σ_n, was determined from $\{\min \mathscr{F}_n\}$. The lowest σ_n value corresponds to the most probable magnitude of n. The confidence interval for n was determined from the upper limit of the 95% confidence interval for σ_n.

On the basis of results obtained, these authors concluded that the maximum confidence interval for n is 2.30–2.75 at various quantum energies (cf. Fig. 4.2).

Later on, the 5/2 power law was verified (without statistical treatment of the results) on a number of metals and in various solutions, including nonaqueous electrolytes (cf., e.g., Rotenberg *et al.*, 1974; Imai 1973). Experimental results obtained for a few solid metals (bismuth, cadmium, indium, lead) are described in the next section. Here, we shall only stress that in all cases hitherto investigated (except for the silver electrode, see below) the 5/2 power law has been confirmed. Any deviations that were observed were connected with conditions deviating from those required by the threshold approximation [Eq. (1.10)] or from one of the conditions formulated on p. 76. Moreover, the observed deviations were utilized themselves for investigation of the actual phenomena which upset the specified conditions (cf. Sections 6.2, 6.5, 7.4, and 8.1).

The dependence of the photocurrent on the frequency (at a constant potential) also obeys a 5/2 power law: thus a straight line is obtained in a plot of $j^{0.4}$ vs. φ as shown in Fig. 4.3 (Korshunov *et al.*, 1971). Figure 4.3 also demonstrates that the coefficient A in Eq. (4.1) depends little (if at all) on the light frequency.

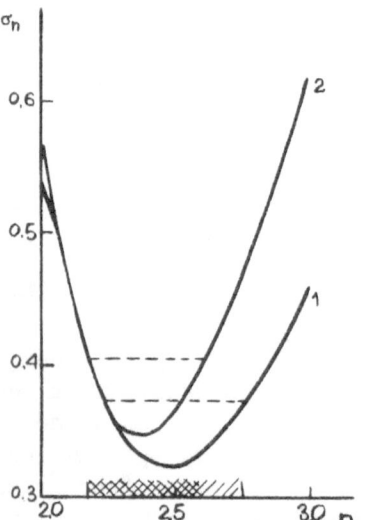

Fig. 4.2. Statistical treatment of the results of the voltametric characteristics of photoemission (Benderskii *et al.*, 1974). Dependence of dispersion on the exponent of the approximating function. Mercury electrode, N_2O-saturated 0.1 *M* KCl. Quantum energy: (1) 3.4 eV; (2) 2.84 eV. Confidence intervals are shown by dashed lines.

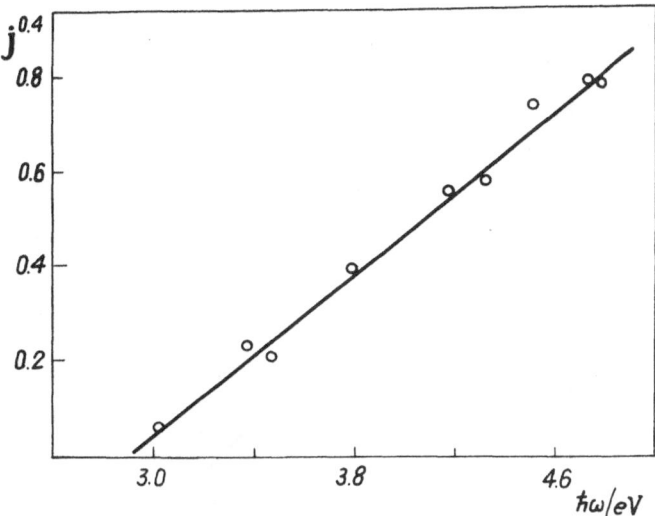

Fig. 4.3. Spectral characteristics of the photoemission current (Korshunov *et al.*, 1971). 0.05 *M* H$_2$SO$_4$ solution, $\varphi = 0.8$ V.

Finally, the dependence of the threshold potential on light frequency has already been mentioned as an important criterion. It can be seen from Fig. 4.4 that a linear plot is indeed obtained with a slope equal to unity (Rotenberg *et al.*, 1974). However, the slope $d(\hbar\omega)/d(e\varphi_0)$ equals unity only in concentrated acceptor solutions, when the photocurrent is close to the emission current. At insufficient acceptor concentrations, the $d(\hbar\omega)/d(e\varphi_0)$ slope may exceed unity, as indicated by De Levie and Kreuser (1969) who worked in N$_2$O-saturated solution.

We shall conclude this section by discussing the quantum yield Y (defined as the ratio of electrons emitted to the number of quanta incident on the electrode surface). The absolute magnitude of Y actually depends considerably on the wavelength and electrode potential. Therefore the quantum yield for various wavelengths should be compared at the same potential, measured with respect to the threshold potential, i.e., at the same maximum energy of emitted electrons: $E_m = \hbar\omega - \hbar\omega_0 - e\varphi = $ const. The quantum yield evaluated for a mercury electrode in 0.5 *N* KF solution saturated with N$_2$O amounts to about 10^{-3} electron/photon when $E_m = 1$ (Rotenberg *et al.*, 1973a). A similar value was obtained by Benderskii *et al.* (1974). For emission currents exceeding 3–4 times the above photocurrent measured in the same solution, the quantum yield is close to 0.003, i.e., similar to that obtained for vacuum emission from the same metals.

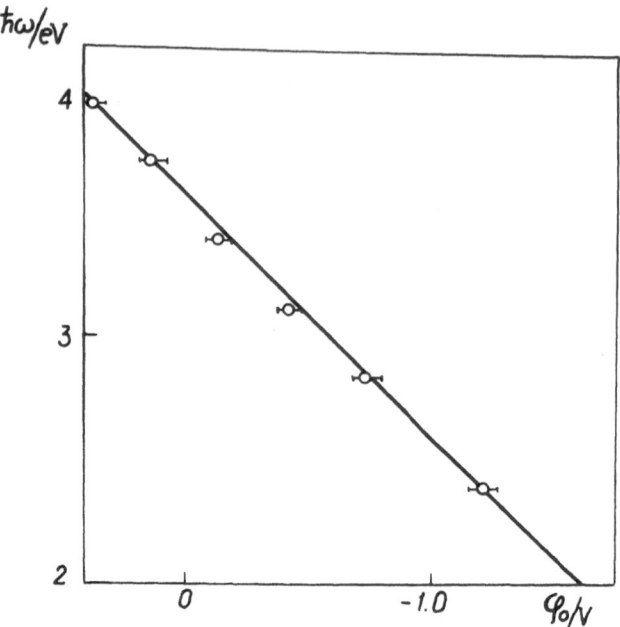

Fig. 4.4. The threshold potential as a function of light frequency (Rotenberg *et al.*, 1974). Lead electrode, 1 *M* HCl.

It is to be noted that the accuracy of quantum yield measurements is relatively low. The quantity Y, however, is not necessary for further considerations, since the photocurrent measurements are usually comparative.

4.3. The Effect of the Nature of the Metal on Photoemission and Electrochemical Kinetics

The photoelectric threshold of photoemission into vacuum depends, as is well known, on the electronic work function of the metal, thus providing a method of measuring the latter. It would appear that by analogy the threshold potential φ_0 at the metal–electrolyte interface should vary with the electronic work function of the metal or, in more accurate terms, with the potential of the zero charge (cf., e.g., Delahay and Srinivasan, 1966).

However, the two interfaces metal–vacuum and metal–solution differ considerably: At the second one a new variable, the electrode potential, is introduced. The metal surface in vacuum is normally uncharged if the metal is not in contact with another, different one. On the contrary, a metal immersed in solution always acquires a certain potential (and charge), which can, in principle, be varied. According to Eq. (0.1) the electronic work function changes with potential, and therefore the work function must be referred to a certain electrode potential. Electrochemical potentials of electrons in metals are the same at the same electrode potential. Therefore, the work of electron transfer from the metal into an electrolyte at *a given constant potential* is *independent* of the nature of the metal and is unambiguously determined by this potential. The specific nature of various metals is reflected only in the height and shape of the potential barrier at the interface (Fig. 4.5). However, the barrier properties are of little importance in the energetics of photoemission under conditions of the threshold approximation. It is solely determined by the difference of the initial and final energy of the electron and does not depend on the potential variations at distances small compared with the de Broglie electron wavelength. Therefore the threshold for photoemission into an electrolyte solution should not depend on the nature of the metal (Rotenberg and Pleskov, 1968; Pleskov and Rotenberg, 1969).

The above statement is valid with respect to the thermodynamic (equilibrium) work function $w^{(th)}$. The latter quantity does not depend on kinetic effects and, in particular, on the law of electron dispersion in the metal. It can be determined, in principle, by measurements of contact

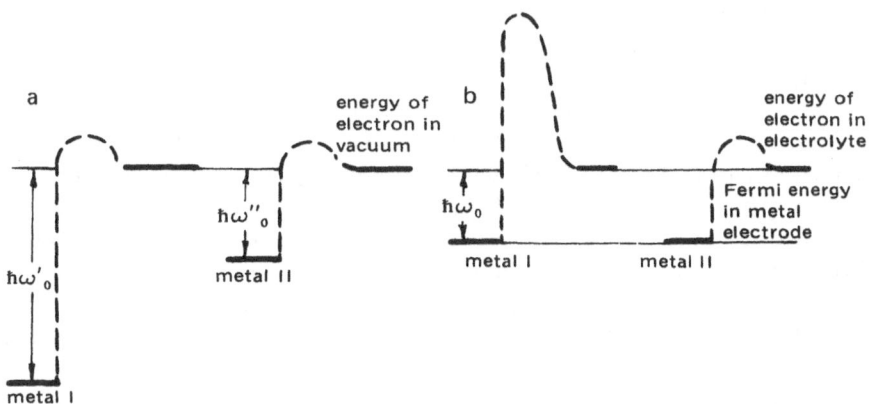

Fig. 4.5. Potential barrier for photoemission at the metal–vacuum (a) and metal–solution (b) interfaces.

potential differences. In emission measurements, however, the electron momentum (quasimomentum) distribution inside the metal plays an important part, resulting sometimes in an "effective" or photoemissive work function, $w^{(ph)}$, being higher than $w^{(th)}$. In fact, the contribution to the photocurrent is made only by electrons with a real value of the square of the x-component of the final momentum, p, i.e.,

$$p^2 = 2m(E_1 + \hbar\omega) - \mathbf{p}_\parallel^2 > 0$$

where E_1 is the initial energy ($E_1 < 0$) and $\mathbf{p}_\parallel^2$ is the square of the momentum (quasimomentum) component, parallel to the metal surface.

For a certain structure of the Fermi surface, the law of electron dispersion in the metal can necessitate emitted electrons exhibiting a tangential momentum different from zero, $\mathbf{p}_\parallel^2 > (\mathbf{p}^{min})^2 > 0$, where $\mathbf{p}_\parallel^{min}$ is the minimum possible value of the momentum $\mathbf{p}_\parallel$. The value of the quantum energy $\hbar\omega = w^{(ph)}$ which ensures, in this case, the positive value of p_x^2 exceeds $w^{(th)}$; the maximum difference for $m = m_0$ is evaluated as 0.1–0.2 eV. The quantities $w^{(ph)}$ and $w^{(th)}$ are equal only if the structure of the Fermi surface contains the point $\mathbf{p}_\parallel^2 = 0$ and is sufficiently smooth around this point. The effective work functions were discussed in detail by Itskovich (1966, 1967) who demonstrated, among others, that investigation of these characteristics of the metal provides one of the methods of studying the shape of the Fermi surface. Gurevich (1969) discussed this problem in connection with electrochemical systems.

The situation described above must be kept in mind in considerations of practical aspects of threshold potential measurements and utilization of results obtained in further calculations. For the majority of metals, and particularly for mercury and liquid amalgams used until now in studies of the effect of the nature of the metal on the threshold potential of photoemission, the required condition of the "smoothness" of the Fermi surface has been satisfied.

The above conclusion concerning the independence of the threshold on the nature of the metal is in agreement with the general laws of electrochemical kinetics. It has been shown by Frumkin (1935, 1965a), Temkin and Frumkin (1955), and Parsons (1964) that the electronic work function (in vacuum) does not explicitly appear in kinetic equations of slow discharge referred to the same electrode potential (with respect to a constant reference electrode). A direct experimental proof of this statement is often difficult to obtain. This is connected, above all, with the breaking of chemical bonds and adsorption of reactants at the electrode surface occurring in the majority of electrochemical reactions, including hydrogen ion discharge. The rate of these side processes depends on the nature of the metal. In the case of simple redox reactions, e.g., anion reduction, the kinetics of the electron transfer reaction in the absence of adsorption was shown to be the same for various metals (Frumkin *et al.*, 1962).

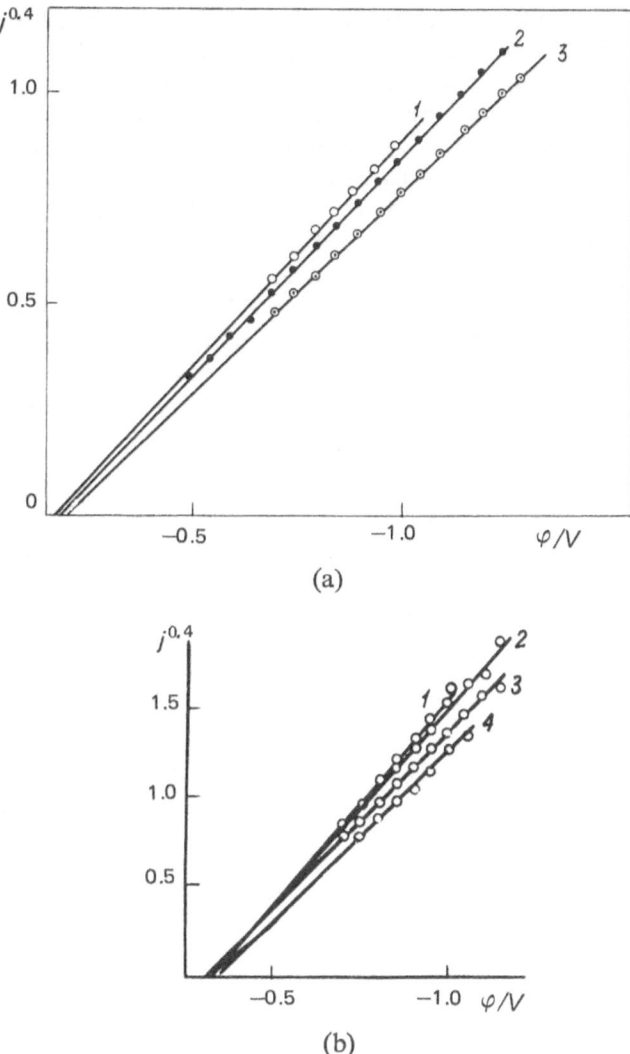

(a)

(b)

Fig. 4.6. The independence of the photoemission threshold on the nature of the metal. (a) Solid metals (Rotenberg *et al.*, 1974). (1) Indium, $0.5\ M$ KF; (2) bismuth, $0.5\ M$ KCl; (3) lead, $0.25\ M$ Na_2SO_4. Acceptor, N_2O. (b) Liquid metals (Rotenberg and Pleskov, 1968). (1) Mercury; (2) thallium amalgam (3%); (3) indium amalgam (18%); (4) indium amalgam (50%); $0.5\ M$ Na_2SO_4 + $0.005\ M$ H_2SO_4. Electrode illumination was somewhat different in various runs.

Photoelectron emission into electrolytes can be considered to be the simplest electrochemical reaction of charge transfer, unaccompanied, as opposed to the majority of electrochemical reactions, by formation or breaking of the bonds. Therefore, it is extremely well suited for elucidation of certain fundamental problems of electrochemical kinetics.

The voltammetric curves, plotted as $j^{0.4}$ vs. φ (Rotenberg et al., 1974), obtained on lead, indium, and bismuth are shown in Fig. 4.6. In spite of large differences (up to 0.5 V according to Leikis et al., 1973) of potentials of zero charge (and consequently of the electronic work functions) of these metals, their threshold potentials do not differ by more than 0.05 V. The independence of the threshold potential on the nature of the metal was confirmed experimentally by Rotenberg and Pleskov (1968) for liquid metals (cf. Fig. 4.6b) and by Zolotovitskii et al. (1972a) on a series of solid metals. A certain analogy to photoemission into vacuum is discussed in Section 10.5.

It must be stressed that the above conclusions are also valid for a particular single electrode, whose potential of zero charge is caused to vary as

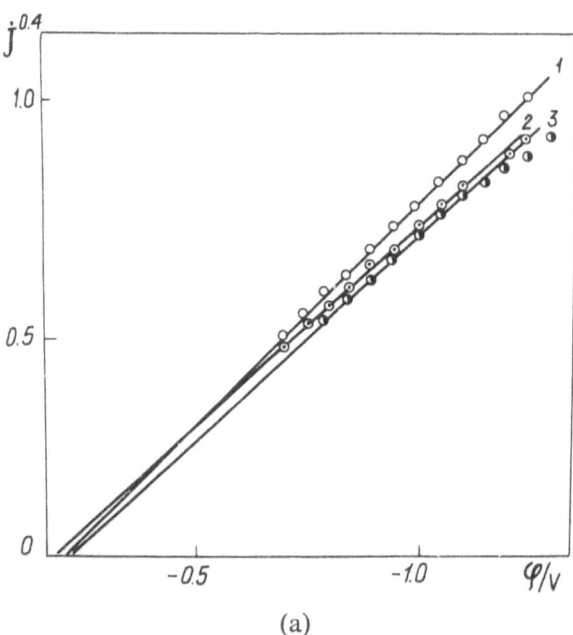

(a)

Fig. 4.7. The independence of the photoemission threshold on specific adsorption. (a) Lead (Rotenberg et al., 1974). (1) 0.25 M Na$_2$SO$_4$; (2) 0.5 M KBr; (3) 0.5 M KI. (b) Bismuth (Rotenberg et al., 1974). (1) 0.5 M KI; (2) 0.5 M KBr; (3) 0.5 M KCl; (4) 0.25 M Na$_2$SO$_4$. (c) Mercury (Rotenberg and Pleskov, 1969). (1) 0.1 M KF; (2) 0.1 M KBr; (3) 0.1 M KI. Acceptor, N$_2$O. Electrode illumination was somewhat different in various runs.

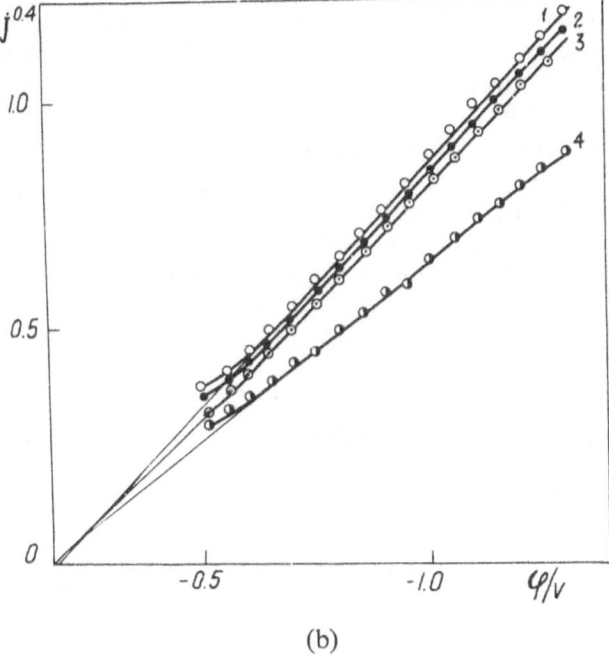

(b)

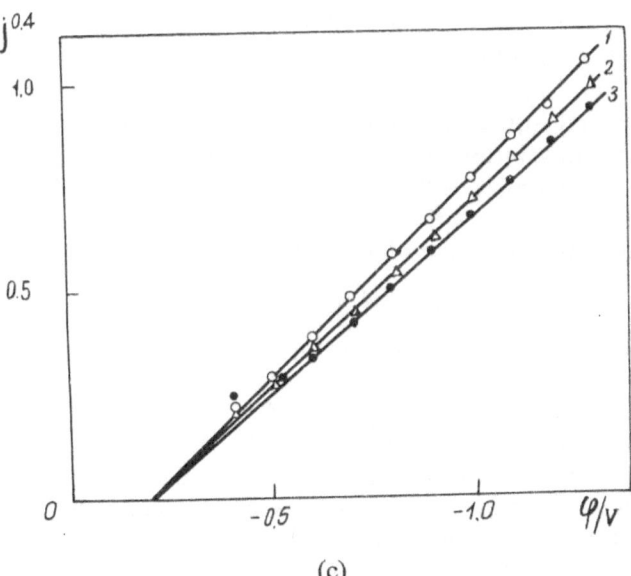

(c)

Fig. 4.7 (continued)

a result of specific adsorption. As long as the "threshold" condition (1.10) holds, i.e., the adsorption layer thickness is less than the de Broglie electron wavelength, the photoemission threshold potential is insensitive to adsorption. The $j^{0.4}$ vs. φ curves obtained for bismuth, lead, and mercury in halide solutions, which strongly affect the potentials of zero charge of these metals, are shown in Fig. 4.7. The analogy between Figs. 4.6 and 4.7 is self-evident and requires no further comments. A similar result was also obtained on a mercury electrode in solutions of certain organic substances (Rotenberg and Pleskov, 1969, 1970). Condition (1.10) does not hold in the presence of thick adsorbed layers, and the experimental curves then do not obey the 5/2 power law (this case is discussed in detail in Section 6.5).

4.4. Energetics of Excess Electrons in Polar Solvents

Photoemission studies, both in vacuum and in solution, allow the energy change in the electron transfer across the interface to be determined.

The electronic work function in solution can be obtained from the relation $w_{ms} = \hbar\omega_0 = \hbar\omega - e\varphi_0$ if the frequency ω and the corresponding threshold potential φ_0 are known. The quantity w_{ms} is referred to an arbitrary electrode potential, chosen as zero on the threshold potential scale. For the given quantum energy, $\hbar\omega - e\varphi_0 = \text{const}$ and, therefore,

$$w_{ms}(\varphi) = e\varphi + \text{const} \tag{4.4}$$

The constant in Eq. (4.4) can be easily found from the $j^{0.4}$ vs. φ plot. In aqueous solutions $\varphi_0 = 0.0 \pm 0.1$ V for the quantum energy 3.4 eV. (This value is obtained for concentrated acceptor solutions when $j \simeq \nu I$ or, after correction, in solutions with low acceptor concentrations; cf. Rotenberg et al., 1970a, b.) Consequently, $\hbar\omega - e\psi_0 = 3.4 \pm 0.1$ eV. Thus, the electronic work function in solution depends on the electrode potential (referred to the saturated calomel electrode) as follows:

$$w_{ms} = 3.4 + e\varphi(\pm 0.1) \text{ eV} \tag{4.5}$$

Values of constants are different in other solvents and for other electrodes. Equation (4.5) allows the work function for emission into solution to be determined at any desired potential. For example, at the potential of the saturated calomel electrode, the work function in water is 3.4 eV; at the potential of zero charge on mercury it is 2.97 eV. Assuming the thermodynamic and photoelectric work functions to be equal (cf. Section 4.3), Eq. (4.5) is valid (for aqueous solutions) independently of the nature of the metal. It can be used, for example, for the establishment of the relationship between the work function in electrolyte and the potential of zero charge. Equation (4.5) does not suffer from the limitations which apply to the well-known Frumkin (1965b) expression for the work function in vacuum in relation to the point of zero charge. (The latter relationship, in principle, may not hold if the orientation of the solvent dipoles at the interface and the deformation of electron clouds at the metal surface depend on the nature of the metal.)

It should be mentioned that photoelectron emission into electrolytes does not make possible the measurement of the absolute potential drop at the metal–solution interface [the problem of absolute potentials has often been discussed, for example, by Ershler (1952), Pleskov and Ershler (1949), and Trasatti (1971)]. In particular, the constant in Eq. (4.5) cannot be identified with the potential difference between the reference electrode and the electron in vacuum, as was done by Zolotovitskii *et al.* (1972b).

Nor can the electron work function in the metal/solution system be found from the "absolute potentials" of Trasatti ε_T (Trasatti, 1974) and Kanevskii ε_K (Kanevskii, 1950). As was shown by Frumkin and Damaskin (1975), ε_T and ε_K are equal to the ideal and to the real free energies of solvation of an electron, respectively, during its transfer into the solution which is in electronic equilibrium with the electrode metal. Unlike ε_T, ε_K can be calculated by a purely thermodynamic method. For a normal hydrogen electrode potential ε_K is equal to 4.44 V.† If we take the potential drop at the solution/air interface to be equal to 0.13 V (Trasatti, 1974), we obtain $\varepsilon_T = 4.31$ eV.

Under conditions corresponding to a hydrated electron concentration of 1 mole-liter^{-1}, $\varepsilon_K = 1.57$ eV (Rotenberg, 1972). From this value we can easily find the value of the standard potential of an "electronic electrode" as referred to the NHE: $1.57 - 4.44 = -2.87$ V.

It is to be noted that the values of ε_T and ε_K are associated with the thermodynamic properties of a hydrated electron, which differs energetically from an electron emitted from metal into solution, the latter being, as has been already pointed out, a quasi-free particle.

On the other hand, photoemission studies allow the evaluation of the interaction energy of an epithermal electron with a polar medium. This requires a comparison of the electronic work function in vacuum, w_{mv}, with the work function in solution, w_{ms}, at the potential of zero charge. The latter value for mercury is close to 3 eV, whereas the work function in vacuum is 4.5 eV. The difference is obviously due to the energy gain resulting from the

---

† This value has been obtained from the equality (Frumkin and Damaskin, 1975) $4.44 = 4.51 - 0.26 + 0.19$ V, where 4.51 is the mercury/vacuum work function, -0.26 is the mercury–aqueous solution Volta potential at the potential of zero charge (Randles 1956), and -0.19 is the potential of zero charge of mercury (NHE).

interaction of the electron with solution. However, the energy of this interaction is not, strictly speaking, equal to $w_{mv} - w_{ms}$. A strict expression can
be derived by considering the contact equilibrium between a metal at its
potential of zero charge, φ_{zc}, and a solution (Fig. 4.8). From the cycle shown,

$$w_{ms}(\varphi_{zc}) = w_{mv} + eV_{ms}(\varphi_{zc}) - U_{sv} \tag{4.6}$$

where V_{ms} is the Volta potential difference between the metal and solution,
and U_{sv} is the interaction energy corresponding to the final state of the
electron in the elementary act of emission. It should not be identified with
the solvation (hydration) energy of the electron (see below). According to
Randles (1956), at the mercury–water interface at φ_{zc}, $U_{sv} = -0.26$ V. Thus

$$U_{sv}^{(H_2O)} = 4.5 - 3.0 - 0.26 \simeq 1.25\,(\pm 0.1)\,\text{eV}$$

Combining the above value with the energy of hydration of the electron,
conclusions can be reached concerning the final state of the emitted electron
in solution.

The free energy of electron hydration is equal to 1.57 eV according to
Rotenberg (1972). This value refers to the standard state (concentration
1 mole-liter^{-1}). The concentration term $-RT \ln (1/55) \simeq 0.1$ eV must, however, be excluded from the latter value since U_{sv}, being a difference of energy
levels, is independent of concentration. (A strict treatment of thermodynamic
relations for the system metal/electrons in solution was given by Frumkin
and Damaskin, 1975, and comparative calculations were given by Conway,
1972.) Comparison of the value obtained (~ 1.5 eV) with the hydration
energy of the electron (1.25 eV) indicates that the potential energy of an
electron emitted into solution exceeds the energy of a hydrated electron by
~ 0.25 eV. The difference cannot be due to the effect of the water orientation
at the interface on the hydration energy, as was suggested by Antropov

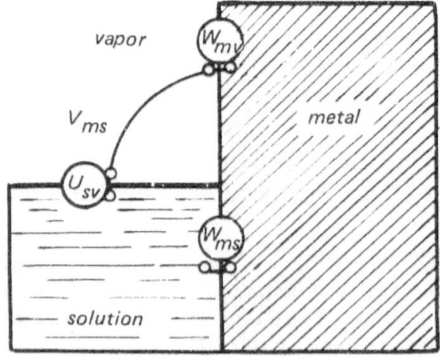

Fig. 4.8. Determination of the interaction energy between the electron and
solution.

(1971). Hydrated electrons are formed at distances exceeding 20 Å from the electrode (see below, Section 5.3) where any solvent-structuring effect of the surface disappears. The latter conclusion is confirmed by the observed independence of the threshold potential on the nature of the metal, regardless of the hydrophilic properties of the electrode, and consequently of the different orientation of water molecules at the surface (e.g., mercury and indium).

The difference between $U_{sv}^{(H_2O)}$ and the total hydration energy of the electron indicates that the final product of the emission step is not a hydrated, but a "dry," electron. This is due to the fact that the emitted electron can be "followed" in time only by the electronic part of the electric polarization of the medium, whereas the solvent molecules do not reorient within the characteristic time of the act of photoemission. The measured value of $U_{sv}^{(H_2O)}$ consequently reflects only this part of the electron hydration energy which is due to the electron's interaction with the "fast" component of the medium polarization only (in particular, with the electronic polarization) and does not include interaction with the "slow" component (e.g., orientation polarization).

Thus, photoemission experiments provide the first (although indirect) evidence of the existence of delocalized electrons in solvents such as water.

The behavior of excess electrons in polar media will now be briefly discussed, owing to the increased interest which this problem has attracted in recent years. The following model-based considerations will be utilized later (Section 9.6) in connection with the energetics of electron photoemission from solution into the gas phase.

The electron energy levels in a polar liquid are illustrated in Fig. 4.9. The reference point (level III) is the energy of the following system: the electron is in the gas phase beyond the range of surface forces; the liquid is in equilibrium (random distribution of solvent dipoles). The value of U_{sv} corresponds to the lower edge of the energy of a dry electron in a liquid, which can be conventionally called (cf. Section 2.1) the bottom of the conduction band (level IV). The quantity U_{sv} consists of the energies of electron polarization, nonelectrostatic interaction of the electron with solvent molecules, and the work of the electron transfer through the potential drop at the solution–vapor interface. The energy of the solvated electron at equilibrium equals its real solvation energy with the opposite sign, $-A_s$. A localized (solvated) electron is in an energy well formed by the given orientation of the solvent molecules. Formation of this well requires an energy E_s (identified, in terms of the simplest model, with the reorganization energy of the medium). Upon absorption of a photon with a sufficient energy, the hydrated electron can be excited to the next level (if it exists), or pass into the conduction band. This transition, being faster than the relaxation

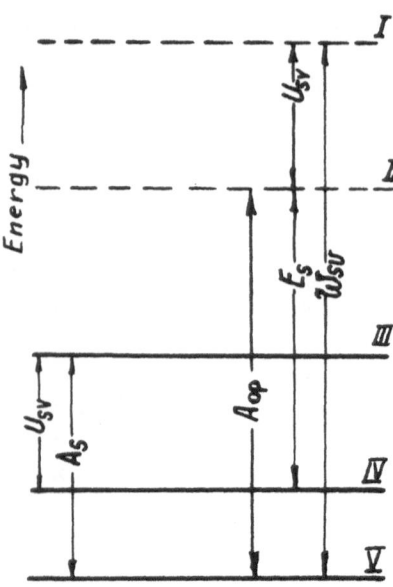

Fig. 4.9. Electron energy levels in solution. (I) Electron in the gas phase, solvent not at equilibrium; (II) delocalized electron in solution, solvent not at equilibrium; (III) electron in the gas phase, solvent at equilibrium; (IV) bottom of the conduction band; (V) solvated electron.

time of the molecules of the medium, does not change the shape of the energy well. Therefore the solvent dipoles retain their initial (spatial and orientational) distribution. The energy of such a transition from the initial localized state of the electron (level V) to level II — photoionization energy A_{op} — can thus, in principle, exceed the energy $A_s - U_{sv}$ by E_s. The electron which has passed into the conduction band (or jumped to an excited level) remains within the liquid phase. The photoemission of a solvated electron from solution thus requires a loss of energy w_{sv}, which exceeds the electron solvation energy by E_s.

Photoemission measurements in nonaqueous solutions have led to the evaluation of the work function of mercury in methanol, ethanol, and di-methylformamide (Zolotovitskii *et al.*, 1972b), as well as in propylene carbonate and hexamethylphosphotriamide (Dogonadze *et al.*, 1974). The differences between the electronic work function of mercury in vacuum and in the above solvents at the potential of zero charge are given in Table 4.1, together with the U_{sv} values calculated from these data. The table contains some other data concerning localized (solvated) and delocalized electrons in polar media, such as solvation energy, A_s; work of electron transfer from solution into the gas phase, w_{sv}; and energy of the maximum in the absorption spectrum of the solvated electron, E_v (the latter quantity allows the energy of transition from the 1s to the 2p level to be calculated). Table 4.1 includes also the Volta potentials of the mercury–solution interface (at the

Table 4.1. Energetics of Excess Electrons in Polar Media (Values in Electron Volts)

Quantity	H_2O	NH_3	Hexamethyl-phosphotri-amide	Methanol	Ethanol	Dimethyl-formamide	Propylene carbonate	Dimethyl sulfoxide
A_s	1.47[a]	1.7[b]	—	—	—	—	—	—
U_{sv}	1.25[c]	—	—	0.9[c]	0.3[q]	−0.1[c]	—	0.07[q]
$(w_{mv} - w_{ms})\varphi_{xo}$	—	—	0.9[c]	0.9[d]	0.75[d]	0.55[d]	1.0[c]	0.74[p]
E_v	1.72[e]	0.88[e]	0.45[f]	1.96[g]	1.77[g]	—	—	—
		1.4 – 1.8[h]	1.34[f]					
w_{sv}		1.42–1.6[i]	1.4[k]					
		1.85[j]						
V_{ms} (Hg, φ_{xo})	−0.26[l]	—	—	−0.53[m]	−0.44[o]	−0.63[n]	—	−0.67[n]
n	1.33	1.33	1.46	1.33	1.36	1.46	1.42	1.47

[a] Rotenberg (1972); [b] Jortner (1959); [c] Dogonadze et al. (1974); [d] Zolotovitskii et al. (1972); [e] Hart and Anbar (1970); [f] Brooks and Dewald (1968); [g] Dorfman (1965); [h] Teal (1947); [i] Hasing (1940); [j] Delahay (1971); [k] Baron et al. (1970); [l] Randles (1956); [m] Case and Parsons (1967); [n] Ganzhina et al. (1971); [o] Damaskin and Kaganovich (1977); [p] Yamashita and Imai (1875); [q] this work.

point of zero charge) and refractive indices of the solvents for visible light frequencies.

For the purposes of the present discussion, the most interesting quantity is the energy of a delocalized electron in the liquid, U_{sv}. Photoemission studies supply a unique possibility for its determination, since photoemission is the only way of introducing low-energy electrons (with energies of the order of 1 eV and less) into a liquid. (Note that the energy U_{sv} obtained by means of photoemission measurements actually describes electron interaction with an ionic solution of sufficiently high ($\simeq 0.1–1$ mole-liter^{-1}) concentration. Therefore it may possibly differ from the energy of electron interaction with pure solvent).

It can be seen from Table 4.1 that U_{sv} varies over a wide range, from 1.25 eV for water to -0.1 eV for dimethylformamide. Similar values of the refractive indices n of the latter solvents indicate that their electronic polarizability should not differ very much. The difference of U_{sv} in various solvents reflects, in the first place, the differences in their surface potentials at the "free" solution interface, χ_s. Secondly, U_{sv} contains a contribution from nonelectrostatic interaction energy between electrons and molecules of the medium, which may differ considerably for various solvents. According to Frumkin *et al.* (1956), χ_s for water amounts to 0.1–0.2 V, the negative end of the dipole pointing outward. Assuming that the surface potential determines largely the value of U_{sv}, the alcohols and dimethylformamide can be concluded to have a positive surface potential, the difference between the surface potentials of water and dimethylformamide being close to 1 V (a similar value, 0.85 V, is reported by Ganzhina *et al.*, 1972).

The difference between the energy levels of a "dry" and solvated electron, $A_s - U_{sv}$, is ~ 0.2 eV for water, and 0.7 eV for hexamethylphosphotriamide (Krishtalik and Alpatova, 1976). Thus, if equilibrium exists between "dry" and solvated electrons, the equilibrium concentration of the former may not be infinitesimally small. In water, for example, the concentration ratio of "dry" to solvated electrons is, as follows from the value of $A_s - U_{sv}$, of the order of 10^{-3}. This can be reflected, for example, in measurements of the reactivity of hydrated electrons, etc.

Zolotovitskii *et al.* (1972b) and Korshunov *et al.* (1971) attempted to make a comparison between the values of U_{sv} resulting from photoemission studies with the theoretically calculated energy of polarons in various media. This comparison, however, was based on a wrong premise: Instead of the U_{sv} value, the authors used the difference $w_{mv} - w_{ms}$, i.e., they did not correct for the Volta potential [cf. Eq. (4.6)]. The latter correction is commensurate with the $w_{mv} - w_{ms}$ value (cf. Table 4.1). Therefore dielectric constants evaluated in the cited papers (necessary for calculations of polaron energy) are not well founded.

4.5. The Effect of the Light Frequency and Polarization. Bulk and Surface Excitation of Electrons

Investigations of frequency and angular dependences of the photo-emission current are needed to gain a deeper insight into photoelectron emission phenomena.

Equation (4.1) brings forward the problem of the frequency dependence of the "pre-threshold" coefficient A. The explicit form of the function $A(\omega)$ depends, according to Section 1.6, on the character of changes of the electromagnetic field of the light wave in the surface layer of the metal. It must be stressed that this dependence is not measurable at the metal–vacuum interface owing to the existence of only a single energy variable, the light frequency. Correspondingly, the change of frequency results in a simultaneous change of both factors in the expression for the photoemission current, namely $A(\omega)$ and $(\omega - \omega_0)^2$. Since the quantity ω_0 is not known *a priori*, the two dependences cannot be separated. Conversely, in electrochemical systems, the change of the factor $(\omega - \omega_0)$ with frequency can always be compensated by a corresponding change of potential, thus making it possible to give an explicit expression for $A(\omega)$. The A values can be found experimentally as the slopes of the $j^{0.4}$ vs. φ plots for each wavelength.

Measurements carried out by Rotenberg *et al.* (1973a) on mercury and lead electrodes in the 2990–4370 Å wavelength range show that A is almost independent of ω (within experimental error). This contradicts the model assuming free penetration of the electromagnetic field in the metal [cf. Eqs. (1.53) and (1.57) for $\mathscr{R} = 0$].

Without going further into the physical interpretation of this result, we shall mention only that the frequency independence of A is borne out experimentally for photoemission in vacuum (cf., e.g., Baumann, 1960): The emission current obeys Fowler's law with high accuracy, assuming independence of the "pre-threshold" coefficient on frequency. In general, all studies of photoemission at the metal–vacuum interface (cf., e.g., Dobretsov and Gomoyunova, 1966) indicate only a very weak frequency dependence of A, and no grounds exist for assuming that a different situation obtains at the interface of the metal with an optically inactive electrolyte.

The experimental frequency dependence of the photocurrent in vacuum is well described by the formula $I \propto (\omega - \omega_0)^2$, which does not allow for a significant frequency dependence of the "pre-threshold" factor. The latter expression is widely used for determinations of the "photoemission" work function $\hbar\omega_0$ (cf., e.g., Rivière, 1969; Fomenko, 1970) and is applicable beyond the slightest doubt for an overwhelming majority of systems investigated.

Korshunov *et al.* (1971) maintain that, for the mercury electrode, A

is strongly frequency dependent ($A \propto \omega^{-4}$). It is impossible, at the moment, to explain this discrepancy of the experimental results. It must be mentioned, however, that the proportionality $A \propto \omega^{-4}$ contradicts the linear dependence of $j^{0.4}$ on $\hbar\omega$ obtained by the same authors in the paper cited (cf. Fig. 4.3).

Following Section 1.1, we can ask which mechanism of photoelectron formation in a given system, bulk or surface excitation, contributes primarily to the photocurrent. The problem has been studied for vacuum systems by Tamm and Shubin, 1931; Mahan, 1970; Spicer, 1968; and Feibelman, 1974. In recent years, electrochemical systems were also investigated in this respect.

Since both the voltametric characteristics and the frequency dependence are the same for the bulk and surface photoeffect (cf., e.g., Benderskii et al., 1974; Babenko et al., 1972), the only way to solve the problem is to investigate the angular and polarization characteristics of photoemission, i.e., the dependence of the photocurrent on the polarization azimuth and angle of incidence. Various aspects of the "vector effect" have been discussed by Görlich, 1962; Farkas et al., 1967; Brauer, 1966; Meessen, 1968; and Fischer, 1972.

According to the surface photoeffect model (cf. Section 1.6), the photocurrent is determined by the square of the normal (to the metal surface) component of the electric vector of the electromagnetic wave, whereas the bulk-excited photocurrent depends primarily on the total luminosity of light absorbed by the metal.

A systematic study of the polarization characteristics of the photocurrent in electrochemical systems was first carried out by Korshunov et al. (1969). The photocurrent j was measured as a function of the polarization azimuth Δ at a constant angle of incidence ϑ. Only a small central part of a mercury pool was illuminated, thus serving as a virtually flat test electrode. The observed dependence of j on Δ obeyed Eq. (1.56) and was interpreted by these authors in terms of the surface photoexcitation effect. Subsequently, however, a number of authors measured the dependence of j on ϑ at constant Δ (for light polarized both in the plane of incidence and perpendicular to the latter) and obtained results which necessitated a revision of the above interpretation.

Typical j vs. ϑ plots are shown in Fig. 4.10. They were obtained by Babenko et al. (1974a, b) at a mercury electrode in KCl solution saturated with N_2O. The authors claim that the character of the curves obtained is independent of the electrolyte, wavelength, electrode potential, and presence in solution of surface-active substances. Qualitatively similar relations were obtained by Barker et al. (1974a).

The authors of the papers quoted have pointed out two important facts.

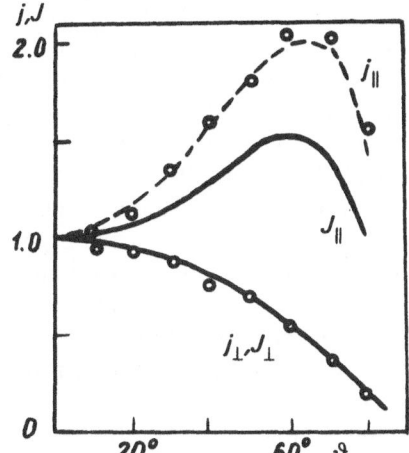

Fig. 4.10. Dependence of the photocurrent *j* and of the luminosity absorbed by the metal *J* on the angle of incidence (Babenko *et al.*, 1974). Light polarized in the plane of incidence (∥) and perpendicular to it (⊥). Mercury electrode, 0.1 *M* KCl solution saturated with N_2O.

1. The photocurrent does not become zero, but has a finite residual value when $\vartheta = 0$. This contradicts the assumption that the photoexcitation originates exclusively from the surface. In fact, in this case, the electric vector lies always (i.e., for nonpolarized and polarized light) in the surface plane, and its normal component equals zero. Correspondingly (within the framework of concepts concerning the surface photoeffect) the light cannot, in this case, excite a photoemission current, assuming that the metal surface is atomically smooth.

It is true that this experimental result must be treated with some caution, since a beam passing from a lamp through a system of lenses is never fully parallel. The error thus introduced is, however, difficult to evaluate accurately.

2. In a number of cases, the angular characteristics of the photoemission current virtually coincide with the angular dependence of the luminosity absorbed by the metal, *J* (the latter quantity was calculated from the Fresnel equations). The *J* vs. ϑ curves are shown in Fig. 4.10 by solid lines. When the electric vector of the light wave is perpendicular to the plane of incidence, the photocurrent $(j_\perp)$ depends on the angle of incidence in the same way as the absorbed luminosity $(J_\perp)$ (cf. Fig. 4.10). The two dependences are different in the case of light polarized parallel to the plane of incidence.

On the basis of the existing experimental material, the following conclusion can be drawn. When the photocurrent is known to be due to light absorption by the bulk electrode, the *j* vs. ϑ and *J* vs. ϑ curves virtually coincide, independently of the type of light polarization. This is usually observed for photoeffects of a nonemissive nature, e.g., at a germanium anode, at metal electrodes covered with semiconducting oxide layers (photocurrent due to the photoconduction effect), as well as on gold and platinum electrodes [photocurrent due to the "photoemission holes"; cf. Section 10.1 (Gerischer *et al.*, 1972)].

On the other hand, when the photocurrent is primarily caused by photoelectron emission, the experimental $j_\perp$ vs. ϑ and $J_\perp$ vs. ϑ curves coincide for light polarized perpendicular to the plane of incidence, whereas the $j_\parallel$ vs. $\vartheta_\parallel$ and $J_\parallel$ vs. $\vartheta_\parallel$ curves obtained for light polarized parallel to the plane of incidence differ, as shown in Fig. 4.10. Both cases are illustrated in Fig. 4.11 for the gold electrode at which, under otherwise the same experimental conditions, anodic polarization results in photoemission of "holes," and cathodic polarization in photoemission of electrons.

Babenko *et al.* (1974a, b) explain this behavior in terms of a considerable contribution to the photocurrent of the surface excitation mechanism which does not lead to a direct relation between the angular polarization characteristics and the energy absorbed by the metal. In other words, the total photoemission quantum yield Y is determined by the bulk Y_b and surface Y_s components as follows:

$$Y = Y_b + Y_s \sin^2 \vartheta$$

The latter authors evaluated the Y_s/Y_b ratio from the curves shown in Fig. 4.10. At the mercury electrode it amounts to 0.4 for a wavelength of

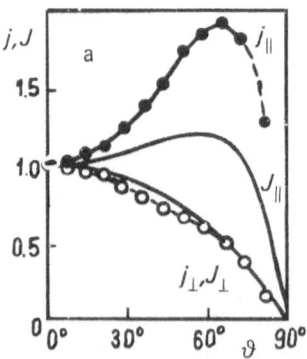
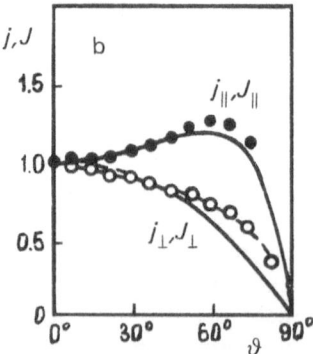

Fig. 4.11.   Photocurrent j and luminosity absorbed by metal J as functions of the angle of incidence (Gerischer, 1973). Light polarized parallel to the plane of incidence ($\parallel$) and perpendicular to the plane of incidence ($\perp$). Gold electrode, 0.5 M H_2SO_4 solution. (a) Cathodic photocurrent (0.1 V); (b) anodic photocurrent (1.4 V). Quantum energy 4.1 eV.

3650 Å. However, the bulk and surface photoexcitation processes are closely linked (resulting, in particular, in the possibility of the appearance of interference terms; cf. Schaich and Ashcroft, 1970). Therefore, when both processes occur with commensurate intensity, the very concept of their quantitative separation does not seem physically valid.

It should be mentioned that the "existence" of the surface photoeffect in the observed photoemission current is supported by the similarity of the potential dependence of the photocurrent and by observation of the same threshold potential, observed at bismuth and antimony electrodes, as that on typical metals (cf. Fig. 4.6a). In this regard, it is to be noted that the band structure of bismuth and antimony qualifies them as semimetals rather than metals, with relatively low concentrations of free electrons; therefore, another dependence of photocurrent on external parameters, in particular the potential, should be expected. Experimental data concerning the double-layer capacity and hydrogen overpotential on Bi and Sb electrodes (cf., e.g., Palm and Tenno, 1973), indicate, however, that in these respects the two elements behave as typical metals.

One of the possible explanations of the absence of significant differences in the electrochemical and photoemission behavior of bismuth and antimony, on the one hand, and other metals, on the other, may consist in assuming a kind of metallization of their surface due, for example, to a high density of surface states. This in turn, may serve as an argument supporting the surface generation (in a thin "metallized" surface layer) of photoelectrons, at least in the base of bismuth and antimony.

Although the existing experimental data and the present state of the theory do not allow final conclusions to be made concerning the nature of photoexcitation, it is obvious now that the photocurrent in electrochemical systems cannot be due to a purely surface, or purely bulk, mechanism of excitation of the emitted electrons. In any case, the two mechanisms do not exhaust all conceivable mechanisms of photoexcitation, in particular when collective interactions are taken into account (cf. Section 1.1). Meanwhile, it has already been mentioned that the surface, or bulk, origin of photoelectrons does not affect the voltammetric characteristics of the photoelectrode. Since the latter characteristics will be primarily utilized in the text following, the actual mechanism(s) of photoelectron generation will not be discussed further.

Finally, it should be mentioned that experiments carried out in polarized light can help to separate photoemission from certain other accompanying photoelectrochemical processes which are insensitive to light polarization (such as bulk photoeffects in the electrolyte, photodecomposition of surface complexes with charge transfer, etc.). This is illustrated in Fig. 4.12 (Korshunov *et al.*, 1969). The voltametric curve (a) was obtained at a

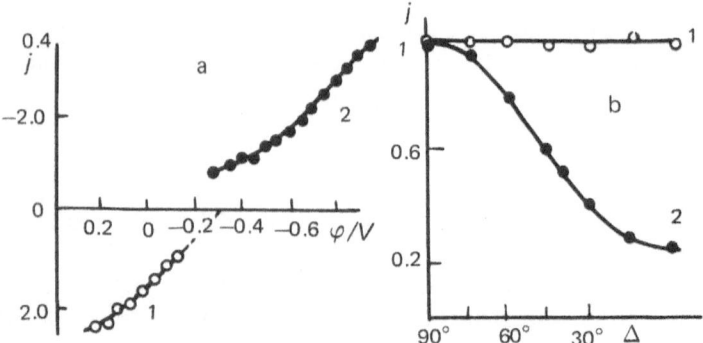

Fig. 4.12. Photocurrent at a mercury electrode in oxalic acid solution (Korshunov *et al.*, 1969). (a) anodic (1) and cathodic (2) photocurrents in 0.05 M $H_2C_2O_4$ as functions of potential; (b) anodic (1) and cathodic (2) photocurrents as a function of the azimuth of polarization of the incident light.

mercury electrode in oxalic acid solution; the light polarization characteristics (b) are shown for the anodic and cathodic parts of the curve. In the cathodic region the photocurrent is due to photoelectron emission (oxalic acid and oxalate anion are acceptors), and it strongly depends on the polarization of the light. Conversely, the anodic photocurrent, being insensitive to the orientation of polarization of light, must be of a nonemissive kind. Korshunov *et al.* (1969) connect this current with the photodecomposition of a surface complex formed by the adsorbed oxalate ion with mercury; Barker and Concialini (1973b) ascribe it to the heterogeneous photolysis of oxalate (see Section 10.1).

4.6. Multiphoton Emission

The experiments described hitherto were concerned with single-photon emission effects: the energy of the light quantum, $\hbar\omega$, being sufficient to increase the energy of the electron, upon absorption of a photon, above the work function in solution, w_{ms}. It was already mentioned in Section 1.1 that photoelectron emission is possible also when $\hbar\omega < w_{ms}$ under conditions of multiphoton absorption, so that $n\hbar\omega > w_{ms} > (n-1)\hbar\omega$, where n is the number of photons absorbed per electron emitted. The probability of the latter process is much less than that of the single-photon emission process (cf., e.g., Brodskii and Gurevich, 1973). Therefore the multiphoton photoeffect can be observed only under conditions of high intensity of the incident light, attainable, for example, with laser sources.

Two- and three-photon photoeffects at the metal–solution interface were first reported by Korshunov *et al.* (1968b).

The dependence of the photoemission intensity of a mercury electrode (illuminated by a ruby laser) on potential is shown in Fig. 4.13 as a plot of $Q^{0.4}$ vs. φ, where $Q = \int j \, dt$ is the charge emitted as a result of the light flash. The curve consists of two sections each approximately obeying the 5/2 power law. Babenko *et al.* (1973) interpreted the curve in terms of a transition from a single-photon photoeffect at potentials below $-1.7\,\text{V}$ to a two-photon effect observed in the range -0.5 to $-1.7\,\text{V}$. In fact, the quantity Q is proportional to the light intensity J in the first region, whereas in the second, a quadratic relation obtains (Fig. 4.14) as predicted theoretically for a two-photon effect [Eq. (1.58)]. The difference of thresholds obtained by extrapolation of the two linear sections in Fig. 4.13 is close to the energy of the ruby laser quantum ($\hbar\omega = 1.78\,\text{eV}$).

The authors stressed the abnormally high ratio of the intensities for two- and single-photon emission (four orders of magnitude higher than that theoretically predicted).

The conclusions of Korshunov *et al.* were criticized by Barker *et al.* (1973). Although the latter authors confirmed some experimental results of Korshunov *et al.* (e.g., the 5/2 power law and the quadratic dependence of the photocurrent on the intensity of the laser light), the dependence of the photocurrent on acceptor concentration was found to be much weaker than that resulting from photoemission accompanied by formation of solvated electrons in solution (cf. Section 2.2). Photocurrents observed in $1\,M$ acceptor solution (HCl) and in the absence of acceptors differed only by a factor of 4 (rather than by two orders of magnitude as for single-photon emission). On this basis, Barker *et al.* (1973) suggested that the photocurrents

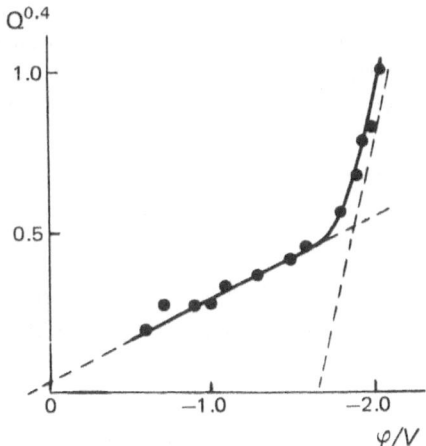

Fig. 4.13. Transition from one- to two-photon emission (Babenko *et al.*, 1973). Mercury electrode, 0.1 M N(C₂H₅)₄NCl solution saturated with N₂O. Ruby laser, $J = 2\,\text{MW-cm}^{-2}$. Dashed line — voltametric characteristics of one- and two-photon emission, calculated from the total emitted charge.

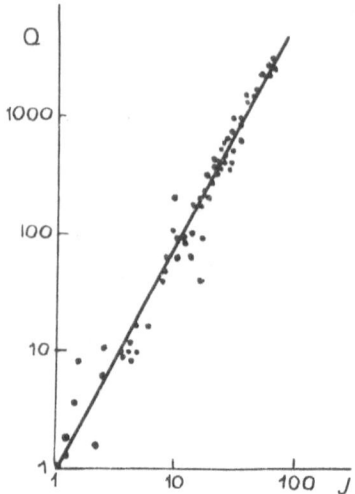

Fig. 4.14. Dependence of charge emitted during two-photon photoemission on the luminosity absorbed by metal (Benderskii *et al.*, 1974). Mercury electrode, 0.1 M KCl solution saturated with N_2O; potential -1.5 V; $J_{max} = 3$ MW-cm^{-2}.

excited by the laser illumination originate not in two-photon emission, but in the interaction of "hot" electrons of the metal electrode with water molecules adsorbed at its surface, or that they may be altogether an artifact due to strong heating of the electrode. In fact, the energy absorbed is in this case of the order of 1 MW-cm^{-2} and the surface temperature may reach 100°C. The "heat-up" currents will be discussed in more detail in Section 10.2. Here it should only be mentioned that they introduce serious experimental difficulties in laser measurements and require special arrangements for separation of photoemission and heat-up signals. Thus, the conclusions of Babenko *et al.* (1973) were put in serious doubt.

Subsequently, Benderskii *et al.* (1974) brought additional arguments in favor of their two-photon emission theory. They claim the luminosity absorbed by the electrode to be extremely large in Barker's experiments (exceeding 3 MW-cm^{-2}). Under these conditions the electrode, in fact, becomes heated up quite strongly, and poorly controlled side effects are thus introduced. At lower absorbed luminosities (0.5–1 MW-cm^{-2}), the dependence of the photocurrent on acceptor concentration, according to Benderskii *et al.*, obeys the laws predicted by the theory of photodiffusion currents.

It can only be concluded that this interesting field of photoemission studies requires further experimental study.

Emitted Electrons in Solution: Subsequent Transformations

The basic laws of photoemission discussed in Chapter 4 did not require consideration of the further behavior of electrons emitted into solution. Therefore, only those experiments were described in which the measured photocurrent was close to the emission current. However, in general, subsequent processes following the act of emission are quite important, especially under conditions of low acceptor concentration. The photocurrent measured under these conditions is lower than the emission current and the basic characteristics of the process (dependence of the photocurrent on potential, light frequency, etc.) can, in principle, differ from the photoemission characteristics.

A sufficiently complete presentation of the photoprocess at the metal–electrolyte interface, including deviations from the photoemission current connected with the return to the electrode of a fraction of the electrons emitted, requires quantitative studies of the consecutive steps which follow the photoemission event.

This chapter is concerned with diffusion processes in aqueous solutions involving hydrated electrons generated in the vicinity of the electrode by the phototransfer of the electron to the aqueous solution. (Data concerning nonaqueous solutions are still unavailable.) The majority of papers on photoemission, starting with the earliest ones (Barker *et al.*, 1966; Delahay and Srinivasan, 1966) and ending with the most recent work (Pleskov and Rotenberg, 1974), involve, to a great extent, this very stage of the process. It is not accidental, since such investigations provide valuable information on the behavior in solution of excess electrons, in particular on their retardation, solvation, and interactions with acceptors and the metal surface. Photoemission plays here the role of a new method of generation and investigation of the properties of hydrated electrons.

In the text which follows (except for Section 5.3), the double-layer thickness is assumed to be small (i.e., concentrated solutions are used) so that the diffusion processes take place beyond its range in the bulk of the solution where charge-neutrality obtains.

5.1. Dependence of the Photocurrent on Acceptor Concentration

One of the criteria differentiating photoemission from other photo-processes at the metal–electrolyte interface is the dependence of the photocurrent j on the acceptor concentration c_A, described [cf. Eq. (2.11)] by

$$j = ev \int_0^\infty \Phi(x)\left[1 - \frac{1}{1 + (Q\mathcal{D}_e/k_s)}\, e^{-Qx}\right] dx \qquad (5.1)$$

where $Q = (k_A c_A/\mathcal{D}_e)^{1/2}$; k_A and k_s are the rate constants for capture of hydrated electrons by the acceptor and electrode surface, respectively; $\mathcal{D}_e$ is the diffusion coefficient of the hydrated electron, and $\Phi(x)$ is the source function for the hydrated electrons.

Equation (5.1) contains, in the general case, a few adjustable parameters (Q, k_s) and one unknown function, $\Phi(x)$, which make its application to analysis of experimental data difficult. In limiting cases, however, Eq. (5.1) can be simplified considerably so that the experimental data can be more easily interpreted.

At low acceptor concentrations, the photocurrent depends linearly on the square root of the acceptor concentration [cf. Eq. (5.1)]:

$$j = v(x_0 + \mathcal{D}_e/k_s)QI \qquad (5.2)$$

where x_0 is the mean distance of retardation and hydration of emitted electrons. In the case of fast capture of hydrated electrons by the surface $(\mathcal{D}_e/k_s \ll x_0)$, we obtain from Eq. (5.2) (substituting for Q):

$$j = vx_0(k_A c_A/\mathcal{D}_e)^{1/2}I \qquad (5.2')$$

For high acceptor concentrations, when $Qx_0 \gg 1$, the photocurrent equals the emission current (within the accuracy of the stoichiometric coefficient)

$$j = vI \qquad (5.3)$$

and is independent of acceptor concentration. Equations (5.2) and (5.3) are more convenient than Eq. (5.1) for quantitative comparison of their predictions with experiment.

In the range $Q\mathcal{D}_e/k_s \ll 1$, i.e., in the case of sufficiently fast electron capture by the surface, Eq. (5.1) reduces to

$$j = ve \int_0^\infty \Phi(x)\{1 - \exp[-(k_A c_A/\mathcal{D}_e)^{1/2}x]\}\, dx \qquad (5.4)$$

The latter equation has been used in many papers (Barker *et al.*, 1966;

Delahay and Srinivasan, 1966; Bomchil *et al.*, 1970; Korshunov *et al.*, 1971) for interpretation of experimentally observed dependences of photo-current on acceptor concentration.

The most suitable acceptor for experimental purposes here is the hydrogen ion. Other acceptors are less convenient owing to their relatively low solubility in aqueous solutions (e.g., in the case of N_2O) or to the potential dependence of the stoichiometric coefficient v (e.g., for the nitrate ion).

A systematic study of the dependence of j on c_A was first carried out by Barker *et al.* (1966) for various acceptors of hydrated electrons. Some of their experimental results obtained on mercury are shown in Fig. 5.1. At low acceptor concentrations, j is indeed proportional to $c_A^{1/2}$; this does not hold, however, at higher acceptor concentrations. Similar relations obtained by Rotenberg (1974) on lead are shown in Fig. 5.2. The photocurrent becomes saturated when sufficiently high concentrations of hydrogen ions are reacted in solution ($c_A > 0.2$ mole-liter^{-1}). A linear dependence of the photocurrent on $c_A^{1/2}$ was also observed at bismuth and indium electrodes (Rotenberg *et al.*, 1975a). Quantitative agreement between the experimental data shown in Figs. 5.1 and 5.2 and the theoretical predictions was the first successful (although indirect) evidence for the existence of photoemission processes in electrochemical systems (Barker *et al.*, 1966).

Although at low acceptor concentrations (where $j \sim c_A^{1/2}$), the experimental data can be unambiguously interpreted in terms of Eq. (5.2), the

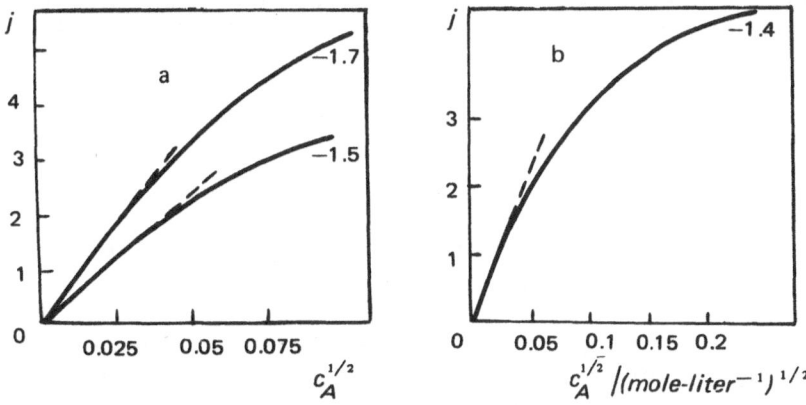

Fig. 5.1. Dependence of the photocurrent on the square root of acceptor concentration (Barker *et al.*, 1966). (a) KNO_3 in 1 *M* KCl; (b) $NaNO_2$ in 0.2 *M* KCl. Mercury electrode, potential (volts) indicated on curves.

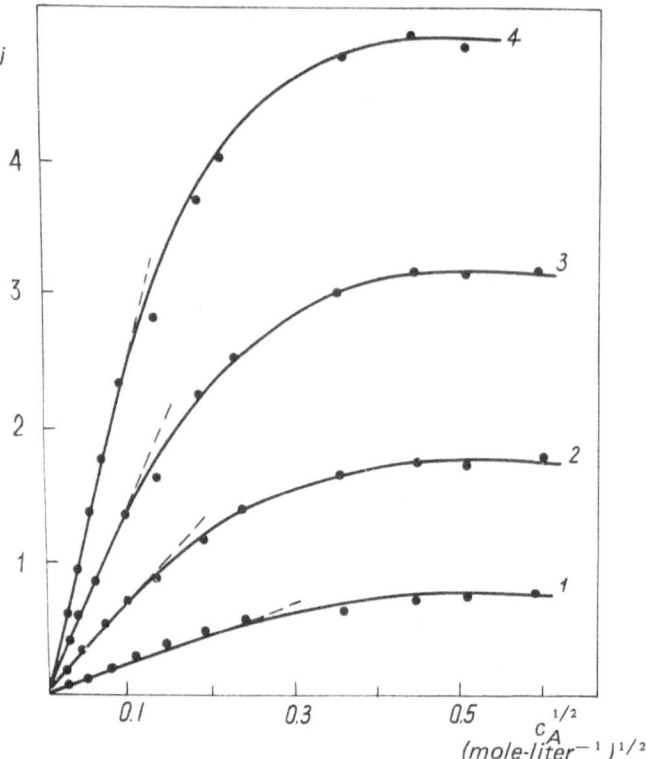

Fig. 5.2. Dependence of the photocurrent on the square root of acceptor concentration (hydrogen ions) (Rotenberg, 1974). Lead electrode; solutions: KCl + HCl (total concentration 1 mole-liter^{-1}). Potentials: (1) -0.7 V; (2) -0.9 V; (3) -1.1 V; (4) -1.3 V.

region of higher concentrations requires additional information concerning the behavior of "dry" electrons in the liquid.

5.2. Is There a Contribution of "Dry" Electrons to Photodiffusion Currents?

A delocalized, dry electron is a predecessor of a solvated electron in the photoemission process. The evidence for this is found, in particular, from studies of, for example, energetics of photoemission as discussed in the preceding chapter. Therefore, the theory of photodiffusion currents must take into account the return of "dry" electrons to the electrode.

The number of these electrons can be evaluated from experimental data on pulse radiolysis of aqueous solutions (Bronskii *et al.*, 1970; Aldrich *et al.*, 1971; Jonah *et al.*, 1973). The solvation kinetics were studied using light pulses of the order of picoseconds. At high (> 1 mole-liter^{-1}) concentrations of acceptors such as NO_3^- or Cd^{2+}, the rate of capture of a dry electron is commensurate with the rate of hydration. Among the acceptors studied only the hydrogen ion virtually does not react with the "dry" electron, although it is extremely effective as an acceptor of solvated electrons.

The difference of reactivity of a "dry" electron with respect to nitrate and hydrogen ions can be used to solve the problem of the possible return of a fraction of "dry" electrons to the electrode before hydration can occur. This requires comparison of photocurrents obtained with both acceptors at a concentration which virtually ensures the capture of all hydrated electrons.

Such a comparison was made by Rotenberg and Gurevich (1975) using a lead electrode in 1 M and 0.1 M solutions of HCl and KNO$_3$. In the range of negative potentials over which secondary reactions can be neglected (cf. Section 8.1), the photocurrent in NO$_3^-$ solution was found to exceed by 1.5 times that observed in the acid solution ($\nu_{NO_3^-} = 3$; see Section 8.1; $\nu_{H_3O^+} = 2$) and, what is most important, this ratio was independent of the acceptor concentration. It follows from this result that "dry" electrons do not return to the electrode, otherwise the ratio of photocurrents observed using the two acceptors should vary considerably with concentration. Thus, Eq. (5.1) can be considered valid.

5.3. Characteristics of Retardation and Hydration of Photoelectrons

Basic Characteristics and Models of the Source Function

The explicit dependence of the photocurrent on acceptor concentration is determined over a wide concentration range by the form of the source function $\Phi(x)$. The function describes the character of retardation and hydration of photoelectrons and is determined, in turn, by the energy distribution of emitted electrons as well as by the energy dependence of the hydration length.

The detailed shape of the function $\Phi(x)$ cannot be derived from the experimental dependence of the photocurrent on acceptor concentration owing to unavoidable experimental errors. The general form of this function,

Table 5.1. First and Second Moments Corresponding
to Various Models of $\Phi(x)$

$\dfrac{e}{I}\,\Phi(x)$	x_0	$\overline{x^2}$	$q = \overline{x^2}/x_0^2$
$\delta(x - x_*)$	x_*	x_*^2	1
$(1/x_*)\theta(x_* - x)^a$	$x_*/2$	$\tfrac{1}{3}x_*^2$	$\tfrac{4}{3}$
$(1/x_*)e^{-x/x_*}$	x_*	$2x_*^2$	2
$(2x/x_*)\theta(x_* - x)$	$\tfrac{2}{3}x_*$	$\tfrac{1}{2}x_*^2$	$\tfrac{9}{8}$

^{a} $\theta(y)$ is the "step" function (see page 22).

however, is quite clear: In the general case $\Phi(x)$ is described by means of its moments, the moment of the kth order being expressed by

$$\overline{x^k} = \frac{e}{I}\int_0^\infty x^k\Phi(x)\,dx \tag{5.5}$$

The first moment x_0 characterizes the "center of gravity" of the $\Phi(x)$ curve with respect to the origin (i.e., the mean hydration length); the second and subsequent moments determine the detailed form of the curve. It is found that at least the first two moments can be rather accurately established from experiment.

Some of the models used in the description of photoelectron hydration are shown in Fig. 5.3. Equation (5.5) allows calculation of the first and second moments for these models (Table 5.1).

All functions shown in the table are standardized in the same way (at unity); however, their first and second moments generally differ. These differences facilitate the choice of the model for $\Phi(x)$ on the basis of experimental determination of both moments.

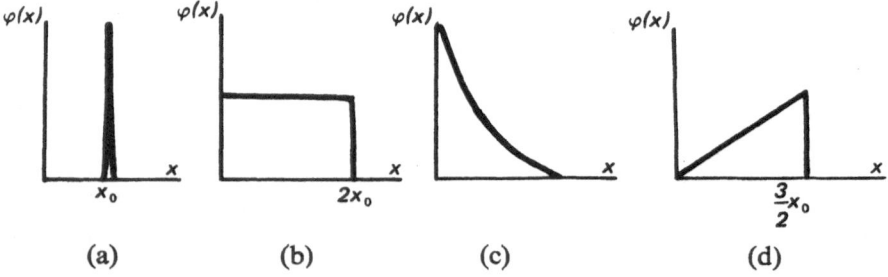

Fig. 5.3. Various models of the source function $\Phi(x)$. (a) $(I/e)\delta(x - x_*)$; (b) $(I/e)\times(1/x_*)\theta(x_* - x)$; (c) $(I/e)(1/x_*)e^{-x/x_*}$; (d) $(I/e)(2x/x_*^2)\theta(x_* - x)$. $\theta(y)$ is the "step" function (see page 22).

Experimental Determination of the Mean Hydration Length [*the First Moment of the* $\Phi(x)$ *Function*]

An approximate estimate of x_0 is based on its relation to the time τ of the (reverse) diffusion of a solvated electron to the electrode: x_0 is of the same order of magnitude as $(\mathscr{D}_e\tau)^{1/2}$. For example, when $20 \lessgtr x_0 \lessgtr 100$ Å, the value of τ is 10^{-8} to 10^{-9} sec. Processes occurring within such short periods of time can only be studied using pulsed light sources with pulses shorter than 10^{-9} sec and an electrical measuring arrangement with a corresponding time resolution. Under these conditions, photopotential (and not photocurrent) is measured, proportional at the given electrode potential to the integral of the photocurrent (cf. Section 3.2). The return of hydrated electrons results in a decrease of photopotential, which tends to zero as $t \to \infty$ when acceptors are absent in the solution and all hydrated electrons are captured by the electrode.

Measurements of the photopotential transients carried out using a nitrogen laser (Korshunov *et al.*, 1971) and the second harmonics of a ruby laser (Barker *et al.*, 1973) with 10^{-8} sec pulse duration were unable to detect the reverse diffusion of hydrated electrons. It follows that the return time must be less than 10^{-8} sec. Therefrom, assuming $\mathscr{D}_e = 5 \cdot 10^{-5}$ cm²-sec⁻¹, $x_0 < 70$ Å.

A more accurate determination of x_0 is based on analysis of the dependence of the photocurrent on acceptor concentration. The problem can be approached in two ways. Barker *et al.* (1966) investigated the concentration dependence of the photocurrent at low and intermediate acceptor concentrations. It is assumed that the capture of hydrated electrons by the surface is a very fast process and $\mathscr{D}_e Q/k_s \ll 1$. (The case of a finite k_s value is discussed in Section 5.4.) Then, taking the linear and quadratic terms of the expansion of Eq. (5.4), we obtain

$$j = \nu I\{x_0(k_A/\mathscr{D}_e)^{1/2}c_A^{1/2} + \tfrac{1}{2}(k_A/\mathscr{D}_e)x_0^2 qc_A\} \tag{5.6}$$

where $q \equiv \overline{x^2}/x_0^2$. Values of q for some models of $\Phi(x)$ are shown in Table 5.1. The slope of the $jc_A^{-1/2}$ vs. $c_A^{1/2}$ plot equals, according to Eq. (5.6),

$$\frac{q}{2}(k_A/\mathscr{D}_e)Ix_0^2$$

The value of x_0 can be found if k_A, $\mathscr{D}_e$, q, and the emission current are known. (The values of k_A and $\mathscr{D}_e$ are usually supplied by radiation-chemical measurements.) Barker *et al.* (1966) obtained for a model (a) (cf. Fig. 5.3) the following values of x_0: 34, 41, and 48 Å at potentials -1.0, -1.2, and -1.4 V, respectively.

It should be noted that the determination of x_0 from Eq. (5.6) does not require a separate measurement of the emission current, as was made in the paper cited. The extrapolation of experimental data plotted as $jc_A^{-1/2} \simeq c_A^{1/2}$ allows the acceptor concentration c_A^* to be calculated from the intercept $jc_A^{-1/2} = 0$ for which $(q/2)x_0(k_A c_A^*/\mathcal{D}_e)^{1/2} = 1$. Thence, x_0 can be found for the given model of $\Phi(x)$ [corresponding to a tabulated (see Table 5.1) value of q] if k_A and $\mathcal{D}_e$ are known.

Another method for determination of x_0 (Gurevich and Rotenberg, 1968) is not model-based and does not use the source function $\Phi(x)$. It utilizes Eq. (5.2) and consists in a comparison of photocurrents observed at low acceptor concentrations (i.e., when j is proportional to $c_A^{1/2}$) with the emission current attainable at high acceptor concentrations.

This method was used for investigating the dependence of the mean hydration length on the external parameters of the system. The dependence of x_0 on the maximum energy of emitted electrons, $E_m = \hbar\omega - \hbar\omega_0 - e\varphi$, is shown in Fig. 5.4. It can be seen that x_0 increases with energy in an approximately linear fashion: $x_0 = a + bE_m$. Expressing x_0 in angstroms and E_m in electron volts, we obtain $a = 5 \pm 2$ Å and $b = 15 \pm 3$ Å/eV. Similar values of x_0 are reported by Pikaev (1970) for the mean path of electrons before hydration takes place, during radiolysis of aqueous solutions.

The increase of x_0 with electron energy is supported by experimental data concerning the dependence of the "apparent" photoemission threshold on acceptor concentration (Rotenberg et al., 1970b) as well as the photoelectron emission at the gas/solution interface (cf. Section 9.6).

Conversely, Korshunov et al. (1970b) consider x_0 to be virtually energy independent. The experimental j vs. $c_A^{1/2}$ curves obtained by the latter authors are the same, independently of the energy of emitted electrons controlled by suitable choice of the light frequency and electrode potential. It is impossible, at the moment, to provide reasons for these experimental discrepancies.

The value of x_0 depends also on the electrolyte concentration (c_{el}). Pleskov et al. (1970) demonstrated that an increase of concentration from 0.5 to 3 M results in a two- to threefold decrease of photocurrent at constant potential, including the potential of zero charge. (Thus, the effect cannot be explained in terms of structural changes of the diffuse layer.) Since, in the case of a neutral acceptor N_2O, the rate constant of electron capture is independent of c_{el} (Pikaev, 1969) and the diffusion coefficient of the hydrated electron should, by analogy with the behavior of other ions, decrease with increasing c_{el}, the observed decrease of the photocurrent must be connected [cf. Eq. (5.2)] with the direct effect of c_{el} on x_0. The values of x_0 reported above refer to ca. 1 M electrolytes.

Bomchil et al. (1970) obtained $x_0 = 50$ Å, using a step-function model for $\Phi(x)$ (second model in Table 5.1). However, the accuracy of the experiments is, according to these authors, insufficient to establish the potential dependence of x_0.

The quantity x_0, shown in Fig. 5.4 as a function of E_m, is averaged over all energies of emitted electrons from 0 to E_m. If the energy distribution of electrons is known, the function $x_0(E_m)$ can be used to determine the functional dependence of the energy of monochromatic electrons, E_f, on their hydration length. It can be shown (Rotenberg, 1974) that for energy distribution described by a power law, the empirical equation $x_0 = a + bE_m$ results in a linear relation between the hydration length and E_f.

Photoelectron emission is a unique method of introducing electrons into liquids with energies of the order of 1 eV. All other methods involve electrons with much higher energies. For example, the literature data concerning thermalization of electrons in aqueous solutions involve electrons with energies exceeding 25 eV (Mozumder and Magee, 1966). Therefore direct comparison of x_0 values obtained from photoemission measurements with literature data is not possible. For the same reason, various theoretical approaches to the problem of retardation of high-energy electrons (e.g., Mozumder and Magee, 1967) can hardly be used for interpretation of the experimental data described above. We can only make the preliminary assumption that in dilute solutions ($c_{el} < 0.1$ mole-liter^{-1}) electrons are

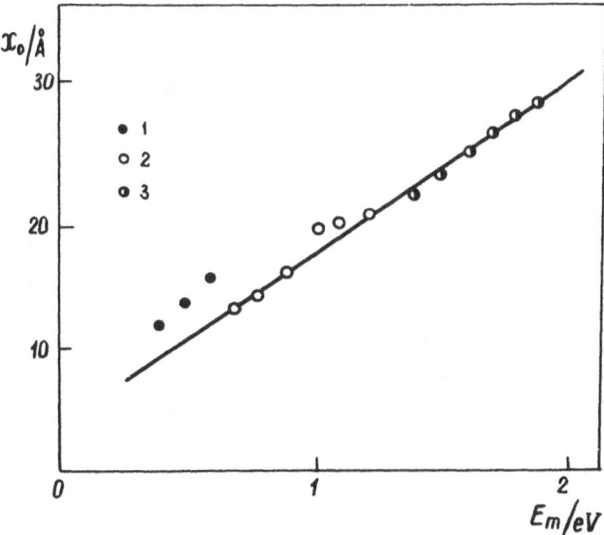

Fig. 5.4. The mean hydration length as a function of the maximum energy of emitted electrons (Rotenberg, 1974). Lead electrode; wavelength: (1) 4370 Å, (2) 3650 Å, (3) 3130 Å.

primarily scattered by water molecules, whereas in concentrated electrolytes electrons are also scattered by ions (Gerischer, 1966b).

Determination of the Second Moment of the Function $\Phi(x)$: Choice of the Model

In spite of the fact that the explicit form of the source function $\Phi(x)$ cannot be established at the presently achievable accuracy of measurements, choice of the proper model for $\Phi(x)$ can be made on the basis of an experimental determination of the second initial moment, $\overline{x^2}$. The latter quantity is related to the first moment, x_0, as follows: $\overline{x^2} = qx_0{}^2$. The coefficient q is found, as described above, by extrapolation of the experimental $jc_A^{-1/2}$ vs. $c_A^{1/2}$ plot. The extrapolated concentration value c_A^* is related to q (see above) as follows:

$$(q/2)x_0(k_A c_A^*/\mathscr{D}_e)^{1/2} = 1$$

The coefficient $(k_A/\mathscr{D}_e)^{1/2}x_0$ can in turn be found by equating the photocurrent extrapolated from the low concentration region, $j = \nu(k_A c_A/\mathscr{D}_e)^{1/2}x_0 I$, to the saturation current νI. The acceptor concentration c^{**} at which both currents are equal can be found from the expression $(k_A c_A^{**}/\mathscr{D}_e)^{1/2}x_0 = 1$. Thence,

$$q = 2(c_A^{**}/c^*)^{1/2}$$

Eq. (5.7) is derived assuming that the electron capture rate by the electrode surface is sufficiently high.

The method described for determination of q is illustrated in Fig. 5.5 with respect to experimental results obtained on mercury, lead, and bismuth in solutions containing hydrogen ions as acceptors of hydrated electrons (Rotenberg et al., 1975a). The $jc_A^{-1/2}$ vs. $c_A^{1/2}$ plots are shown there, together with the dependence of j on $c_A^{1/2}$ and with the saturation currents (straight lines) measured in $1\ M$ HCl solution. It can be seen from Fig. 5.5 that the values of c_A^* and c_A^{**} are virtually the same for all three metals, i.e., the nature of the metal does not affect the values of x_0 and $\overline{x^2}$. This, in turn, confirms the validity of the form chosen for the function $\Phi(x)$.

The average values of q for mercury, lead, and bismuth are 1.1, 1.15, and 1.15, respectively. They correspond best (cf. Table 5.1) to the (triangular) model shown in Fig. 5.3d. Satisfactory agreement is also obtained using the model shown in Fig. 5.3b. Other models enumerated in Table 5.1 are not in agreement with experiment.

Independent information concerning the shape of $\Phi(x)$ can be obtained by changing the concentration of an indifferent electrolyte (i.e., by studying the effect of the ψ' potential on the photocurrent). Photocurrents calculated

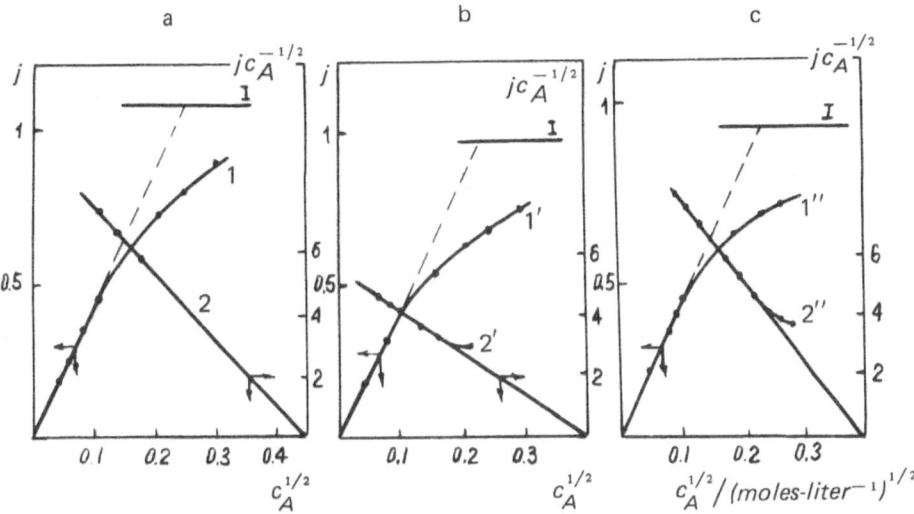

Fig. 5.5. Determination of the second moment of the source function $\Phi(x)$ (Rotenberg *et al.*, 1975). Plots of j vs. $c_A^{1/2}$ and $jc_A^{-1/2}$ vs. $c_A^{1/2}$ for mercury (a), lead (b), and bismuth (c) electrodes in acid solutions at -0.1 V.

from Eq. (2.18) for three models of $\Phi(x)$ (delta function, exponential function, and step function; cf. Fig. 5.3) are shown in Fig. 5.6 as functions of the electrolyte concentration c_{el} at $\psi' < 0$ (Rotenberg and Gurevich, 1973). The numerical value of the ψ' potential was chosen for each concentration far from the point of zero charge, where ψ' depends little (logarithmically) on the electrode potential φ. The parameter m was chosen for the actual system studied (saturated N_2O solution).

Various theoretical curves in Fig. 5.6 (solid lines) correspond to different values of the hydration length x_0, which appears here as a parameter. The type of dependence of j on c_{el} is rather sensitive to the form of the source function $\Phi(x)$, thus enabling (by comparison of calculation with experiment) an adequate model to be chosen. Experimental data are shown in Fig. 5.6 by the dashed lines. The effect of the ψ' potential on photoemission itself was evaluated by comparing experimental photocurrents at the same potential, after correcting for the ψ' potential. (The correction was made only for 0.01 and 0.001 M solutions; at higher concentrations the diffuse layer is transparent to emitted electrons; cf. Section 6.2.) Experimental and theoretical dependences are normalized to coincide at $c_{el} = 0.5\ M$ (for several arbitrary x_0 values).

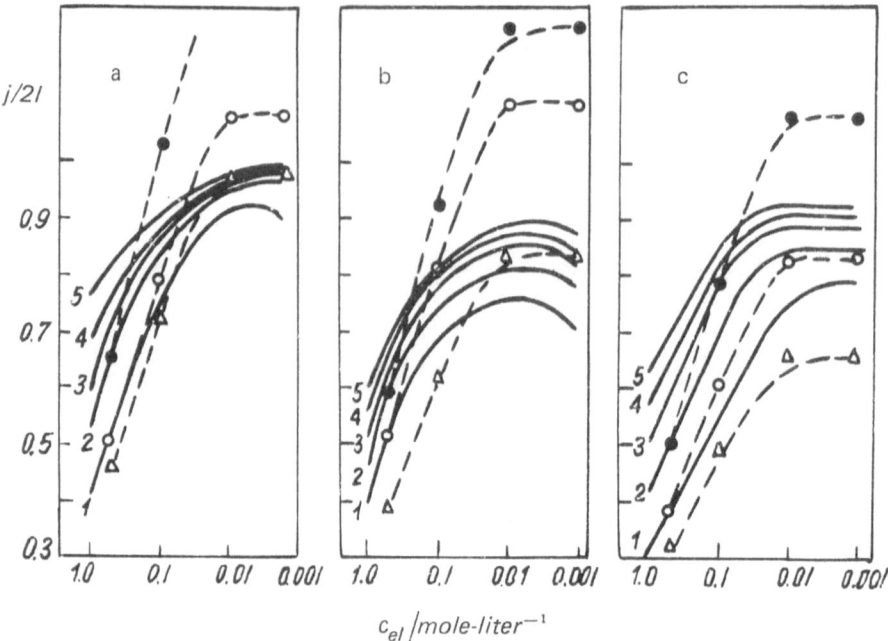

Fig. 5.6.   Effect of the ψ' potential on photodiffusion currents calculated from the dependence of $j/2I$ on electrolyte concentration (Rotenberg and Gurevich, 1973). Source function: (a) $(I/e)\delta(x - x_*)$; (b) $(I/e)(1/x_*)e^{-x/x_*}$; (c) $(I/e)(1/x_*)\theta(x_* - x)$. Hydration length x_0: (1) 20 Å, (2) 30 Å, (3) 40 Å, (4) 50 Å, (5) 60 Å. Dashed lines are drawn through experimental points (KF solutions saturated with N_2O) and standardized with theoretical ones (●, for $x_0 = 30$ Å; ○, for $x_0 = 20$ Å; △, for $x_0 = 15$ Å). Corresponding theoretical curves are not shown. Lead electrode; wavelength: (1) 4370 Å, (2) 3650 Å, (3) 3130 Å.

It can be seen from Fig. 5.6 that the experimental points do not follow any of the lines in the calculated family of curves. This can be caused by the increase of mean hydration length x_0 with electrolyte dilution. The experimental data for 1 M solution agree best with the curves calculated for the step-function model with $x_0 = 20$ Å. The triangular model (Fig. 5.3d) was not discussed in the paper cited.

5.4. Slow Capture of Hydrated Electrons by the Metal Surface

The capture of hydrated electrons by the metal surface should be, from the thermodynamic viewpoint, very facile. However, kinetic inhibitions can, under certain conditions, slow down the process to relatively low rates.

The possibility of slow capture of hydrated electrons was first suggested by Barker *et al.* (1966). These authors assumed capture to be preceded by a dehydration step

$$e_{aq}^- \rightleftharpoons e^- + H_2O$$

which may control the overall rate of the process. However, no experimental data supporting this hypothesis were reported.

The capture rate can be evaluated from Eq. (2.13). Thus, for low acceptor concentrations, when $Qx_0 \ll 1$,

$$j = \nu \frac{x_0 + (\mathscr{D}_e/k_s)}{1 + (Q\mathscr{D}_e/k_s)} QI \tag{5.8}$$

The deviation of the j vs. $c_A^{1/2}$ plots from straight lines occurs the earlier (i.e., at lower acceptor concentrations) the lower is k_s. The magnitude of k_s for metals such as mercury and lead is of the order 10^3 cm-sec^{-1} and higher. Results of experiments carried out at the indium electrode are shown in Fig. 5.7 (Rotenberg and Pleskov, 1973), together with those obtained at mercury. The scales of the two plots are chosen so as to make them coincide at low values of the concentration. At low concentrations of the acceptor (hydrogen ion; $c_A < 0.001\ M$), both plots are linear. At higher concentrations both photocurrents deviate from the linear dependence toward lower values; the deviation occurs earlier for indium than for the mercury electrode. Since all constants in Eq. (5.8), except for k_s, are independent of the nature

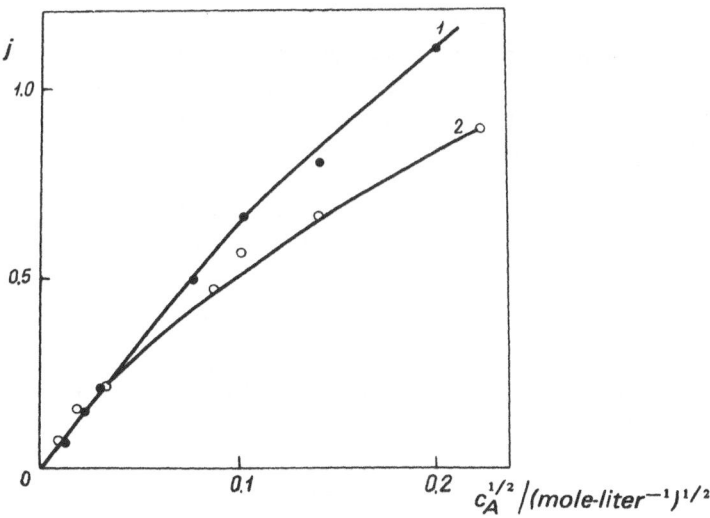

Fig. 5.7. Dependence of photocurrent on $c_A^{1/2}$ (Rotenberg and Pleskov, 1973). (1) Mercury electrode; (2) indium electrode. Acceptor, hydrogen ions; potential, -0.1 V.

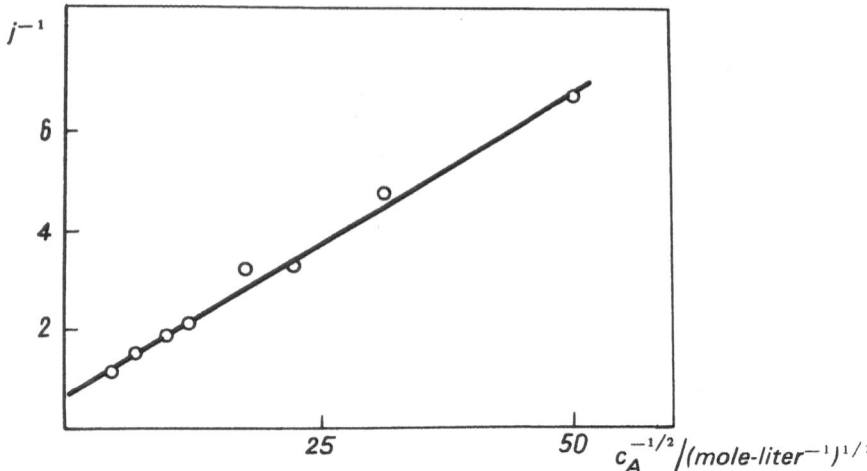

Fig. 5.8. Experimental data for indium electrode replotted (from Fig. 5.7) as j^{-1} vs. $c_A^{-1/2}$ (Rotenberg and Pleskov, 1973).

of the metal, the observed effect can be ascribed to the inhibited capture of hydrated electrons by the indium surface. In this case, at intermediate concentrations, for which the inequality $Qx_0 \ll l$ still holds, a linear j^{-1} vs. $c_A^{-1/2}$ behavior should be expected. This is in fact observed experimentally, as shown in Fig. 5.8. The value of k_s evaluated from this plot, using literature values for k_A and $\mathscr{D}_e$, is 100 cm-sec^{-1}.

The lower (as compared to mercury, lead, bismuth) rate constant for electron capture observed at the indium electrode can be connected with the presence of an oxide film which cannot be removed even under cathodic polarization, or with the specific adsorption of water molecules as indicated by capacity measurements (Grigor'ev et al., 1972). Whatever the case, the structural properties of the indium–electrolyte interface inhibit the penetration of electrons from the outer Helmholtz plane directly to the metal surface.

5.5. Measurements of the Rate Constants of Electron Capture by Acceptors

The values of the kinetic parameters k_A and $\mathscr{D}_e$ necessary for calculations of x_0 have been based usually on independent measurements carried out primarily under conditions of pulse radiolysis. However, radiation-chemical measurements are usually carried out in very dilute electrolyte solutions. Therefore, the values of k_A and $\mathscr{D}_e$ obtained in this way may, generally speaking, be inapplicable to the more concentrated solutions used in measurements of photoemission currents.

The ratio $k_A/\mathscr{D}_e$ can be directly evaluated, although with a rather low accuracy, from measurements of photodiffusion currents observed under conditions of acceptor reduction at the electrode. From Eqs. (2.20) and (5.2), the ratio of the photocurrent obtained in the absence of acceptor discharge, j, and in the presence of the latter, j_*, is given by

$$j/j_* = (2/3^{1/3})(k_A c_A/\mathscr{D}_e)^{1/6}\delta_N^{1/3} \tag{5.9}$$

Measurements of j and j_* were carried out by Rotenberg and Gurevich (1975).

Their results obtained at lead and bismuth electrodes in 0.5 M KCl + 0.001 M HCl solution are shown in Fig. 5.9. Owing to the high hydrogen overpotential on lead, H_3O^+ ions are not discharged on this metal in the potential range studied. Therefore the vicinity of the cathode is not depleted in hydrogen ions and the photocurrent changes monotonically, whereas at the bismuth electrode hydrogen ions are discharged at sufficiently negative potentials. The limiting current (i.e., $c_{H_3O^+} \simeq 0$) is reached at $\varphi = -1.4$, a point corresponding to a minimum on the photocurrent curve. Both curves in Fig. 5.9 are constructed to coincide in the region of more positive potentials. The photocurrent observed at the lead electrode can thus be considered

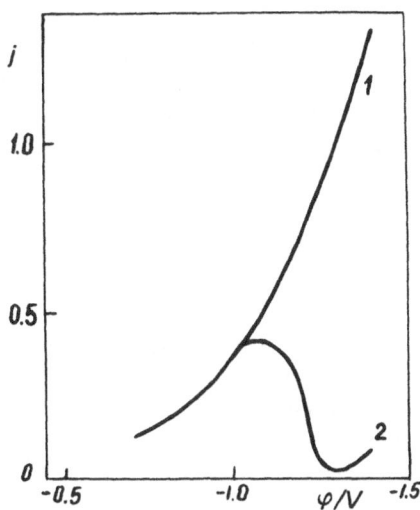

Fig. 5.9. Photocurrents on lead (1) and bismuth (2) as functions of potential (Rotenberg and Gurevich, 1975); 0.5 M KCl + 0.001 M HCl.

equal to that which would be obtained on bismuth in the absence of hydrogen ion discharge. The diffuse-layer thickness δ_N can be calculated from the limiting current for hydrogen ion discharge (in the dark) if the diffusion coefficient of H_3O^+ ions is known (cf., e.g., Pleskov and Filinovskii, 1976). The value of $k_A/\mathscr{D}_e$ found in this way [Eq. (5.9)] equals 10^{14} to 10^{15} cm^{-2}-mole^{-1}-liter and that resulting from pulse radiolysis, $4 \cdot 10^{14}$ cm^{-2}-mole^{-1}-liter. The photoemission method is apparently accurate to within one order of magnitude.

The absolute values of k_A can be determined using nonstationary photoemission methods. The previously mentioned method of periodic (harmonic) electrode illumination provides the possibility of determining kinetic constants from the phase shift between the measured photocurrent and the light intensity. Experimental solution of this problem requires relatively high frequencies of light modulation.

Barker (1971) and Barker et al. (1974b, c) used 10^{-6} to 10^{-8} sec pulses; they compared the photopotential $\varphi_{ph}^{(1)}$ observed after a fixed period of time t in the absence of electron acceptors with the corresponding photopotential $\varphi_{ph}^{(2)}$ observed in the presence of the acceptor (e.g., N_2O) after periods much shorter than t. Barker (1971) expresses the ratio of both photopotentials by

$$\varphi_{ph}^{(1)}/\varphi_{ph}^{(2)} = \tfrac{1}{2}(\pi k_A c_A t)^{-1/2}$$

from which k_A is readily recovered. The rate constant of electron capture by N_2O in 1 M KCl solution was found to be $4 \cdot 10^9$ mole^{-1}-liter-sec^{-1}, in good agreement with literature (radiation chemical) data. Barker et al. (1974b) consider this result to be another confirmation of nonparticipation of "dry" electrons in the reaction with acceptors at low concentration of the latter.

The pulse illumination method can also be used for other purposes than determination of rate constants for hydrated electron capture. Thus, Barker (1971) and Zolotovitskii et al. (1975) utilized the method to evaluate [cf. Eq. (2.37)] the diffusion coefficients of H˙ and OH˙ radicals in aqueous solutions.

5.6. Method of Competing Acceptors in Photoemission Studies

The relative values of rate constants of electron capture by acceptors can very often be determined in radiation-chemical measurements by means of the competing acceptor method (cf., e.g., Pikaev, 1969). Barker et al. (1966) and later Rotenberg et al. (1972) used this method to investigate homogeneous reactions of hydrated electrons which follow photoelectron emission.

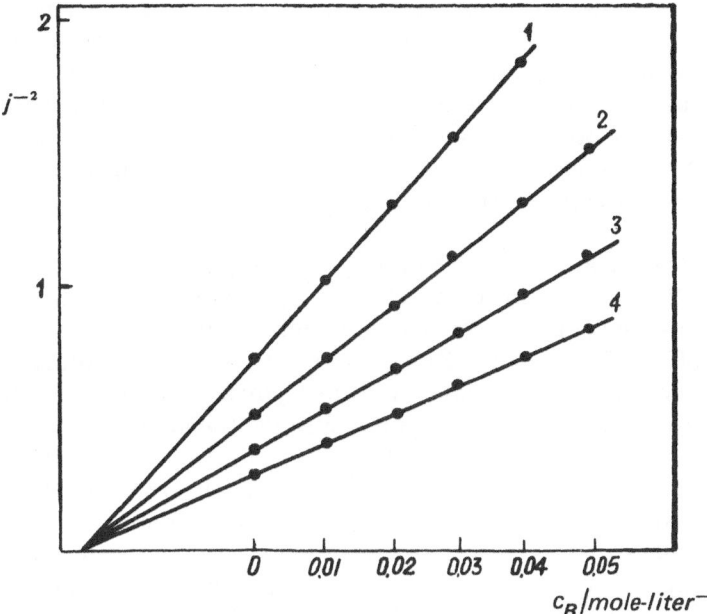

Fig. 5.10. Determination of the relative rate constants for hydrated electron capture using the competing acceptor method (Rotenberg *et al.*, 1972). 0.7 *M* NaF solution saturated with N_2O with addition of acetone. Potential: -1.0 (1); -1.1 (2); -1.2 (3); -1.4 (4) V.

The measurements are carried out in a solution containing two acceptors, A and B; after the electron capture, one of them, A, is reduced; the other, B, is oxidized at the electrode. Obviously, the photocurrent decreases with increasing relative concentration of B. The relation between the photocurrent j and the concentrations of the acceptors A and B was derived in Section 2.2.

The limiting expression (2.29a)

$$j = \frac{2k_A c_A x_0}{\mathscr{D}_e^{1/2}(k_A c_A + k_B c_B)^{1/2}} I \qquad (5.10)$$

is most convenient for experimental purposes. Equation (5.10) is valid under conditions $x_0(k_A c_A + k_B c_B)^{1/2} \mathscr{D}_e^{-1/2} \ll 1$, which are relatively easy to satisfy experimentally. They require concentrated electrolyte solutions (x_0 lower than in dilute ones) with low acceptor concentrations. According to Eq. (5.10), j^{-2} should be a linear function of c_B.

The method is illustrated by a j^{-2} vs. c_B plot (Fig. 5.10) obtained in NaF solution saturated with N_2O (acceptor A). The concentration of acetone (acceptor B) did not exceed 0.05 *M*. The experimental points satisfactorily follow a straight line. Extrapolation of the experimental lines to $j^{-2} = 0$ leads to a value of k_B/k_A which is close to 1, in good agreement with the results of radiation-chemical measurements (cf. Pikaev, 1969). Similar results are obtained using other acceptor pairs (Barker *et al.*, 1966).

The absolute values of rate constants for electron capture are given in Table 5.2 for a number of acceptor pairs used in photoemission and pulse radiolysis measurements. It can be seen that both methods yield similar results.

The rate constants of electron capture by divalent metal ions are included in Table 5.2. The capture of an electron reduces them to univalent ions, which are usually unstable both in the bulk solution and on the surface. For example, in Pb^{2+}, Cd^{2+}, Zn^{2+}, and Cu^{2+} solutions, photocurrents are negligibly small over the whole accessible potential range (i.e., from the threshold potential to that corresponding to the potential of "dark" reduction of the divalent ion at the electrode). Obviously, the univalent ions of the metals studied undergo fast oxidation at the electrode. Although Zn^{2+} and Cd^{2+} do not yield their "own" photocurrents, the rate constants for electron capture by these acceptors can still be measured using the method of competing acceptors described above (which requires the capture product [eA], in this case univalent Zn^+ or Cd^+ ions, to become oxidized at the electrode).

Conversely, considerable photocurrents are observed in Ni^{2+} and Mn^{2+} solutions; this indicates a possibility of inhibited oxidation of Ni^+ and Mn^+ ions in the potential range studied (Barker *et al.*, 1966).

The results discussed in Chapter 5 show that experimentally measured photocurrents depend not only on the magnitude of the photoemission current in solution, but also on processes connected with solvation and

Table 5.2.   Rate Constants for Electron Capture by Acceptors

Acceptor pairs		Rate constant $k_B \cdot 10^{-10}$, mole^{-1}-liter-sec^{-1}	
A	B	Photoemission	Pulse radiolysis[c]
N_2O	H_3O^+	1.5[a]	1.4
H_3O^+	Zn^{2+}	0.046[a]	0.06
N_2O	Cd^{2+}	3.6[a]	2.1
N_2O	Ni^{2+}	1.5[a]	0.9
N_2O	NO_3^-	2.0[a]	1.7
N_2O	NO_2^-	0.5[a]	0.68
N_2O	Acetone	0.6[b]	0.59
N_2O	Thiourea	0.4[b]	0.3

[a] Barker *et al.* (1966); [b] Rotenberg *et al.* (1972); [c] Hart and Anbar (1970).

subsequent transformations of the emitted electrons. Therefore, measurements carried out at low acceptor concentrations are less suitable for studies of the photoemission process itself. Meanwhile, changing acceptor concentration in a wide range, simultaneous utilization of two and more acceptors, changing concentration of the indifferent electrolyte, comparison of data obtained with charged and neutral acceptors, as well as use of pulse and modulated illumination techniques, provide a wide range of possibilities for studying the kinetic characteristics of homogeneous reactions of solvated electrons by photoemission methods.

Investigation of the Structure of the Electric Double Layer Using the Photoemission Method

The processes initiated by photoemission take place in a region near the electrodes, overlapping that of the double layer. Therefore, the structure of the interface should affect considerably both the photoemission itself and the processes connected with formation and subsequent transformations of solvated electrons.

6.1. The Role of the Diffuse Layer in Photoemission Phenomena; General Considerations

The effect of the diffuse layer structure on photoelectron emission can be treated by analogy with its effect on the kinetics of electrode reactions. The key role of the double layer in electrochemical kinetics was first pointed out by Frumkin (1933) in his interpretation of the experimental data concerning hydrogen overpotential. Quantitatively, the effect of the structure of the double layer on electrode kinetics is accounted for by introduction of the ψ' potential into the kinetic equation for slow discharge of ions. The effect of the ψ' potential is twofold. First, it changes the energy levels of the initial and final state of the system and, consequently, the activation energy and the electrochemical rate constant of the process. Second, it affects the local concentration of charged reactants in the vicinity of the electrode.

The dependence of the ψ' potential on the surface charge and electrolyte composition is described in the self-consistent field approximation by the Gouy–Chapman theory (cf., e.g., Parsons, 1954; Delahay, 1965).

The first experimental verification of Frumkin's theory was reported by Levina and Zarinskii (1937) who investigated hydrogen ion discharge at Hg electrodes in solutions of various compositions. Subsequently, the range of systems studied was increased considerably, in particular, after the anomalous polarographic behavior of anions in dilute solutions had been observed by Kryukova (1949). At present, an extensive body of experimental material on double-layer effects on kinetics exists, involving, in particular, reduction of anions at various metals (Nikolaeva-Fedorovich, 1970; Frumkin *et al.*, 1959; Petry and Nikolaeva-Fedorovich, 1961), which fully supports Frumkin's theory.

Electron photoemission can be treated as the simplest electrochemical reaction. Therefore, double-layer effects should, in a certain sense, be similar in photoemission phenomena and electrochemical kinetics. They are, essentially, as listed below.

1. The double-layer structure affects the emission process itself, i.e., the elementary act of electron transfer from the metal into the electrolyte. It has been shown theoretically in Chapter 1, and experimentally in Chapter 4, that under conditions of the threshold approximation, when the double-layer thickness does not exceed the De Broglie wavelength of the electron the structure of the interface practically does not affect the photoemission behavior. However, in dilute solutions the diffuse layer thickness becomes greater than the electron wavelength. The electron cannot then penetrate through the potential barrier due to the diffuse layer, so that the structure of the latter (quantitatively expressed by the magnitude of the ψ' potential) then affects considerably the rate of photoemission. At negative surface charges, the emission current decreases compared with that in concentrated solutions, since the electron requires an additional energy $|e\psi'|$ to overcome the potential barrier of the diffuse layer (Fig. 1.2a). Conversely, at positive surface charges, the electron can gain some energy upon transfer into solution (Fig. 1.2b).

2. The double-layer structure affects the following secondary processes accompanying photoemission: (a) migration in the electric field of solvated electrons returning to the electrode (cf. Section 2.2); (b) interaction of electrons with charged acceptors, whose distribution at the interface depends on the double-layer structure; and (c) migration of charged capture products [eA] (e.g., NO_3^{2-}).

3. The rate constants of heterogeneous processes appear in the expression for the photocurrent if the electrode reactions of the capture products [eA] are kinetically rather than diffusion controlled, or if the capture of the solvated electron by the electrode surface is inhibited. These rate constants depend, generally speaking, on the ψ' potential and thus the double-layer structure affects the resulting photocurrent.

The effect of the double-layer structure on the elementary emission act becomes significant when the double-layer thickness exceeds the wavelength of the emitted electron. The photodiffusion processes are affected if the diffuse-layer thickness is commensurate or exceeds the thermalization and solvation length. Since the latter distance is of the order of 20–30 Å (cf. Section 5.3), a significant diffuse-layer effect on photodiffusion processes can be expected only in dilute electrolyte solutions.

Table 6.1.   Effect of the ψ' Potential on the Observed Photocurrent

ψ' Potential effects	Change of the photocurrent (as compared to conditions when $\psi' = 0$)	
	at $\psi' < 0$	at $\psi' > 0$
Photoemission itself	decreases	remains the same or increases
Migration of solvated electrons to the electrode surface	increases	decreases
Concentration at the interface of		
(a) neutral acceptor	does not change	
(b) positively charged acceptor	increases	decreases
(c) negatively charged acceptor	decreases	increases
Migration of a negatively charged capture product [eA]		
(a) with subsequent electro-oxidation	increases	decreases
(b) with subsequent electroreduction	decreases	increases
Migration of a positively charged capture product [eA]		
(a) with subsequent electro-oxidation	decreases	increases
(b) with subsequent electroreduction	increases	decreases
Rate of electrode reactions of the capture products [eA]	may increase or decrease	

Possible mechanisms of the effect of the diffuse double-layer field on the overall photoelectrochemical process are shown in Table 6.1.

Depending on the system studied, the overall effect of the double layer, consisting of the factors enumerated in Table 6.1, may lead either to a decrease or increase of the photocurrent as compared with the result when $\psi' = 0$.

Finally, it must be taken into account that a change of the electrolyte concentration can affect the photocurrent not only directly owing to the change of the ψ' potential, but also owing to the change of the ionic strength of the solution which can affect the rate of the homogeneous interaction of the solvated electron with a charged acceptor (Hart and Anbar, 1970), as well as the solvation length x_0 (cf. Section 5.3).

The effect of the diffuse-layer structure on the act of electron emission itself for $\psi' < 0$ depends quantitatively on the numerical value of the parameter $(2me|\psi'|)^{1/2}/\kappa\hbar$ (cf. Section 1.5), which characterizes the ratio of the effective diffuse-layer thickness κ^{-1} to the de Broglie wavelength of the emitted electron with the energy $e|\psi'|$.

When

$$(2me|\psi'|)^{1/2}/\kappa\hbar \gg 1 \tag{6.1}$$

the diffuse layer becomes nontransparent to emitted electrons, i.e., electrons with energies less than $e|\psi'|$ cannot penetrate the barrier. Quantitative evaluations show that inequality (6.1) holds only in relatively dilute solutions ($c_{el} < 0.1$ mole-liter^{-1}). In more concentrated solutions, electrons freely tunnel through the potential barrier and the ψ' potential does not then significantly affect photoemission.

At negative ψ' potential values, the emission current in dilute solutions is given by (cf. Section 1.5)

$$I = A[\hbar\omega - \hbar\omega_0 - e(\varphi - \psi')]^{5/2} \qquad (6.2)$$

which differs from Eq. (4.1) in that φ is replaced by $\varphi - \psi'$. If $\psi' > 0$, the potential barrier due to the ψ' potential is replaced by a potential well (Fig. 1.2b). The applicability of Eq. (6.2) is not well founded in this case. However, it will be shown below that owing to a number of additional reasons, Eq. (6.2) also remains valid under certain conditions when $\psi' > 0$.

Equation (6.2) contains a "ψ' correction" similar to that which appears in the equations of electrochemical kinetics (Frumkin *et al.*, 1952). In both cases (photoemission and kinetics of electrode reactions) the rate of the electron transfer through the interface is affected not by the full metal–solution potential drop, but by the part $(\varphi - \psi')$. However, it must be taken into account that the effective, one-dimensional potential in Eq. (6.2) which can be determined experimentally using the photoemission method (see below) is not identical with the ψ' potential in the kinetic equation for electrode reactions introduced in order to account for the "real" micropotential effect on ion discharge kinetics treated in classical terms (cf. p. 30).

The effect of the ψ' potential on migration of solvated electrons was discussed in Sections 2.2 and 5.3.

6.2. Dependence of the Photocurrent on Electrolyte Concentration

In the absence of specific adsorption of ions and molecules, the structure of the double layer is primarily determined by the electrolyte concentration c_{el}. It affects not only the value of the ψ' potential but also the process of thermalization of emitted electrons. In principle, the two effects can be separated by comparison of photocurrents at the potential of zero charge, i.e., in the absence of the ψ' effect. More difficult to separate are effects of the ψ' potential on the emission itself and on photodiffusion currents. For this purpose, the electrolyte should contain uncharged acceptors, which do not change the ionic strength of the solution and whose distribution at the interface is independent of potential. The simplest system suitable for separation of the two effects is an electrolyte saturated with N_2O. The double-layer structure affects in this case only the emission step and the return of hydrated electrons to the electrode. The theoretical analysis of both phenomena (Sections 1.5 and 2.2) simplifies considerably the interpretation of experimental data.

The effect of the electrolyte concentration is shown in Fig. 6.1 (mercury electrode) for KF. At negative surface charges, the photocurrent decreases with c_{el} in the series $0.01 > 0.001 > 0.1 > 0.5 M$ solutions. For 0.001 and 0.01 M solutions, the curves intersect at the potential of zero charge ($\varphi_{zc} = 0.43$ V). Therefore, in this concentration range, dilution affects only the structure of the double layer. Curves obtained in 0.1 and, particularly, in 0.5 M solutions intersect the curves obtained in dilute solutions at potentials more positive than φ_{zc}. This indicates that the further increase of the electrolyte concentration results in an additional decrease of j, connected only in part with the change of the double-layer structure, being primarily due to a decrease of the mean solvation length of photoelectrons at high electrolyte concentrations (cf. Section 5.3).

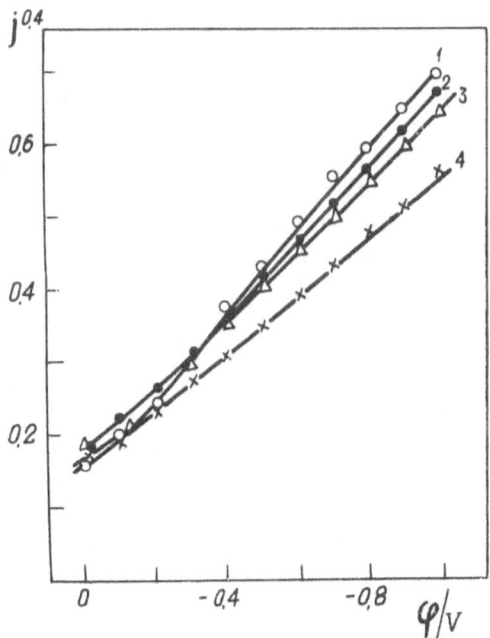

Fig. 6.1. Effect of the electrolyte concentration on the photocurrent: $j^{0.4}$ vs. φ curves obtained at the mercury electrode in KF solutions saturated with N_2O (Rotenberg *et al.*, 1973). KF concentration (mole-liter^{-1}): (1) 0.01; (2) 0.001; (3) 0.1; (4) 0.5. Wavelength 2990 Å.

The decrease of photocurrent with dilution in the range of low ($c_{el} <$ 0.01 M) concentrations is connected, as indicated qualitatively by Eq. (6.2), with the effect of the ψ' potential on emission. Moreover, it can be seen from Fig. 5.6 that electron migration in the diffuse-layer field is virtually independent of electrolyte concentration in this range. This should be ascribed to the compensation of the increase of $|\psi'|$ potential with dilution by a corresponding increase of the Debye screening length, κ^{-1}. As a result of this compensating feature, the effective field acting on returning electrons varies little with the electrolyte concentration.

On the other hand, the same calculation (Fig. 5.6) shows that at negative surface charges the photocurrent in dilute solutions is close to the emission current and, therefore, photodiffusion processes do not affect the resulting photocurrent. In the electrolyte concentration range 0.01 to 0.5 M, the primary effect is that of the ψ' potential on the migration of hydrated electrons.

6.3. Measurement of the Potential of Zero Charge by a Photoemission Method

At the potential of zero charge, and in the absence of specific adsorption of molecules and ions, $\psi' = 0$ and all effects connected with the diffuse-layer field disappear. Therefore, in the low concentration range the j vs. φ curves intersect at φ_{zc} (cf. Fig. 6.1).

The same situation should be expected in a more complex case, when the double-layer structure affects not only the emission step and the electron migration, as in N_2O solutions, but also, for example, the concentration of charged acceptors at the electrode. The photoemission method for determining the potentials of zero charge is based on the latter case.

The $j^{0.4}$ vs. φ curves obtained in acid solutions at various concentrations of the background electrolyte at lead, cadmium, and indium electrodes are shown in Fig. 6.2. The same type of curves is observed at bismuth and antimony electrodes. In the negative surface charge range, the photocurrent on all these metals decreases monotonically with increasing total concentration of the electrolyte. This behavior differs from that shown in Fig. 6.1 for N_2O, indicating that the ψ' potential affects primarily photodiffusion processes, being of secondary importance in the emission act itself [cf. Eq. (6.2)]. The observed dependence of j on c_{el} is in qualitative agreement with the numerical calculations for a charged acceptor, H_3O^+, carried out by Bomchil et al. (1970).

It can be concluded from Figs. 6.1 and 6.2 that the potential of zero charge is a special point on the photocurrent curve, independent of the

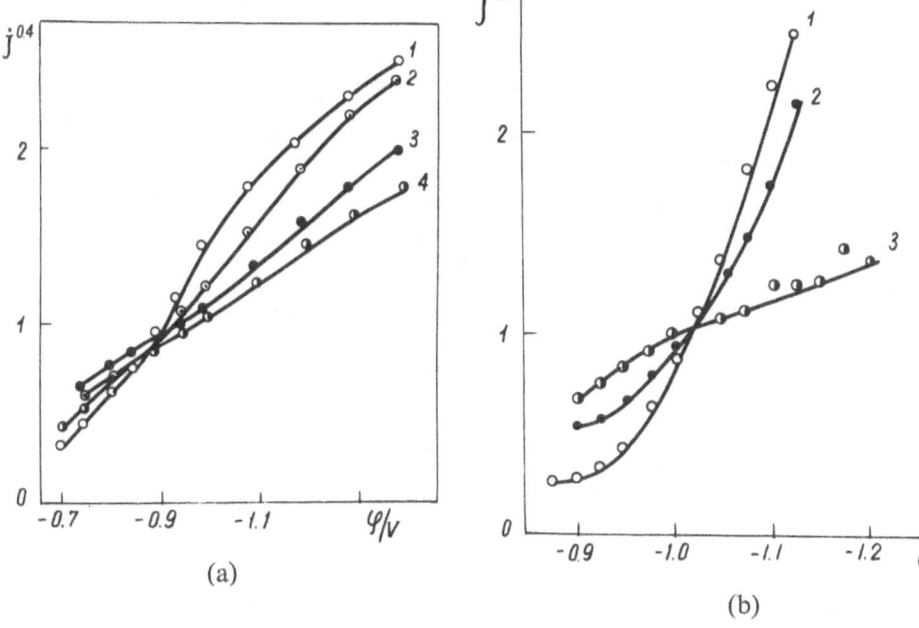

(a)

(b)

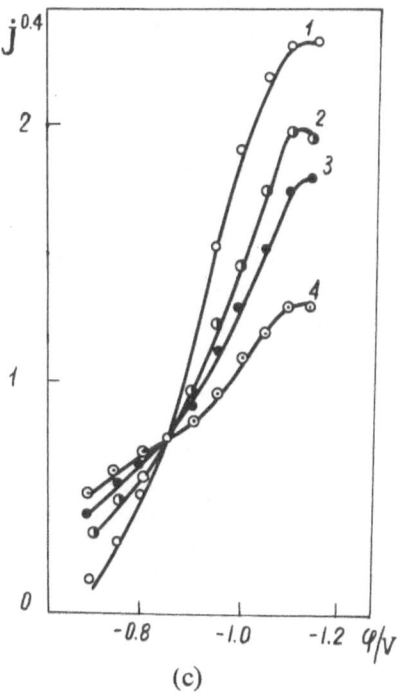

Fig. 6.2.   Determination of the potential of zero charge. Method I — from the intersection point of photocurrent–potential curves (Rotenberg *et al.*, 1974). (a) lead, 0.001 *M* HCl solution with addition of NaCl [(1) 0; (2) 0.01; (3) 0.1; (4) 0.23 *M*]; (b) cadmium, 0.001 *M* HCl with KCl addition [(1) 0; (2) 0.01; (3) 0.1 *M*]; (c) indium, 0.001 *M* HCl solution with addition of KCl [(1) 0; (2) 0.01; (3) 0.03; (4) 0.1 *M*].

(c)

Table 6.2. Potentials of Zero Charge

| Metal | φ_{zc} (V) | |
	Photoemission	Differential capacity[d]
Mercury	-0.43[a]	-0.43
Lead	-0.85[b]	-0.81
Bismuth	-0.63[b]	-0.64
Cadmium	-1.02[b]	-1.0
Indium	-0.90[b]	-0.90
Antimony	-0.40[c]	-0.40

[a] Pleskov and Rotenberg (1969); [b] Rotenberg *et al.*
(1974); [c] Rotenberg *et al.* (1975b); [d] Frumkin (1972),
Leikis *et al.* (1973).

wavelength. It can be best seen in dilute electrolyte solutions. The intersection point of photocurrent curves obtained at various electrolyte concentrations and constant acceptor concentration is then a quantitative
measure of the potential of zero charge. The values of the potential of zero
charge obtained using the photoemission method are compared in Table 6.2
with data calculated from capacity measurements. The tabulated data show
that both methods yield quantitatively similar results. The second variant
of the photoemission method is discussed in the next section.

6.4. Direct Experimental Determination of the ψ' Potential

Insofar as the explicit dependence of the photoemission current on the
ψ' potential is known [Eq. (6.2)], comparison of experimental data obtained
in dilute electrolyte solutions with those obtained at the same acceptor
concentration in concentrated solutions ($\psi' \neq 0$) enables, in principle, the
ψ' potential to be determined. This requires, however, that all other effects
[enumerated in Section 6.1 but unaccounted for by Eq. (6.2)] be excluded.
Comparison of Eqs. (6.2) and (4.1) shows that the photocurrent vs.
potential curve in dilute solutions is shifted with respect to a similar curve in
more concentrated solutions by the value of the ψ' potential.
In order to avoid complications connected with the effect of the ψ'
potential on photodiffusion processes, the comparison must involve limiting
emission currents. These, however, can be experimentally observed only at
high ($\sim 1\ M$) concentrations of sufficiently effective acceptors. The solubility
of the most suitable neutral acceptor, N_2O, amounts at atmospheric pressure

to only 0.025 mole-liter^{-1}. Thus, the saturation current in this case can be reached only at pressures of several tens of atmospheres. In the presence of high concentrations of charged acceptors (e.g., H_3O^+ or NO_3^-) the diffuse layer practically disappears. Consequently a direct experimental determination of the ψ' potential is rather difficult to make by means of photoemission measurements under normal conditions.

Nevertheless the problem can be solved, at least over a limited range of concentrations and potentials. Thus, as was already mentioned in Section 5.4, the diffuse-layer field effectively hinders the return of solvated electrons to the electrode at sufficiently negative ψ' values (i.e., in dilute solutions at potentials much more negative than φ_{zc}). The limiting photocurrent can be reached under these conditions at lower acceptor concentrations than in concentrated solutions in which $\psi' = 0$ (cf., e.g., Fig. 5.6).

The experimental data presented below allow the ψ' potentials to be determined quite accurately in relative terms as a function of the electrode potential in dilute electrolytes and in absolute terms for negatively charged metal surfaces. Thus, the potential difference observed at the same photocurrent j for 0.001 and 0.1 M KF solutions (cf. Figs. 6.1 and 5.6) is practically equal to the difference of their ψ' potentials. The difference, $\Delta\psi'$, found in this way is plotted in Fig. 6.3 as a function of the electrode potential together

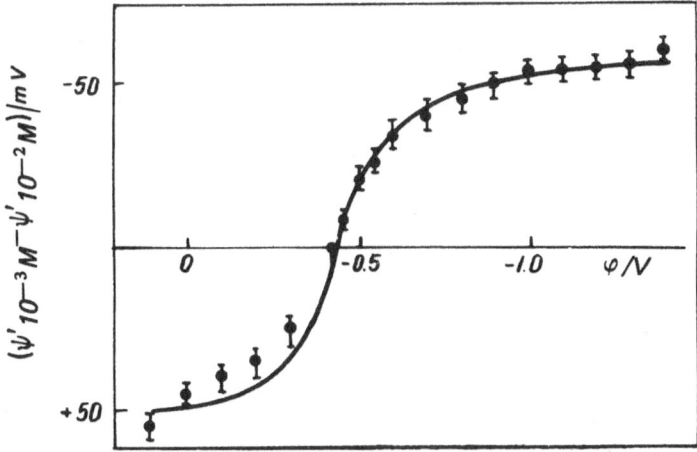

Fig. 6.3. The difference of ψ' potentials in 10^{-3} and 10^{-2} M KF solutions as a function of the electrode potential (Rotenberg *et al.*, 1973b). Solid line calculated from the Gouy–Chapman theory.

with a curve calculated from the Gouy–Chapman theory. The good agreement between the experimental and calculated curves confirms that the structure of the double layer in dilute electrolytes at intermediate acceptor concentration (N_2O-saturated solution) affects primarily the photoemission step itself, i.e., Eq. (6.2) is quite strictly obeyed. The change of the sign of the effect at the potential of zero charge indicates the validity of this equation at positive ψ' values as well. It can be concluded that in dilute solutions, when $\psi' > 0$, electrons can pass from the metal through the energy well at the interface (Fig. 1.2b). This transfer can evidently occur if the well thickness (the effective diffuse-layer thickness) exceeds both the de Broglie wavelength of emitted electrons and the mean thermalization and solvation length. Both conditions are well known to be satisfied in dilute solutions.

It can be seen from Fig. 6.3 that the ψ' potential changes are most pronounced at the point of zero charge. The ψ' vs. φ curve has an inflection at this point. Differentiation of Eq. (6.2) with respect to $-\varphi$ (assuming A to be φ-independent) leads to

$$d(I^{0.4})/d(-\varphi) = (eA)^{0.4}(1 - d\psi'/d\varphi)$$

Since the derivative $d\psi'/d\varphi$ has a maximum at φ_{zc}, the derivative $d(I^{0.4})/d(-\varphi)$, as well as (approximately) $d(j^{0.4})/d(-\varphi)$, should exhibit a minimum. The latter is used for evaluation of the potential of zero charge. The dependence $d(j^{0.4})/d(-\varphi)$ vs. φ is shown in Fig. 6.4 for 0.01 M KF solution saturated with N_2O. The potential corresponding to the minimum on the curve is close to the φ_{zc} value obtained by the method previously described.

The limiting emission current can be attained at sufficiently negative

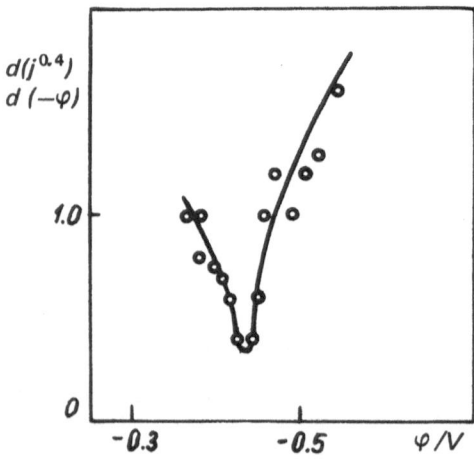

Fig. 6.4.   Determination of the potential of zero charge. Method II — from the minimum on the $d(j^{0.4})/d(-\varphi)$ vs. φ curve (Pleskov et al., 1970). Mercury electrode, 0.01 KCl solution saturated with N_2O.

potentials in dilute electrolytes at much lower acceptor concentrations than in concentrated solution, where $\psi' \simeq 0$. This effect is particularly pronounced in the case of cationic acceptors (e.g., hydrogen ions) whose concentration at the interface exceeds that in the bulk. Under these conditions, the ψ' potential can be absolutely measured by comparison of the j vs. φ curves obtained in dilute and concentrated electrolytes. In both cases, the photo-current is virtually equal to the emission current. The ψ' potentials thus measured are shown as a function of the electrolyte concentration (in 0.1–0.001 M acid solutions) in Fig. 6.5. Good agreement with the Gouy–Chapman theory is found in this case as well (Prishchepa *et al.*, 1975).

It should be stressed once more that the photoemission method does not lead to the determination of the local ψ' potential, but to the mean value of the potential in the Helmholtz plane which appears in the Gouy–Chapman theory. The photoemission method, based on the quantum-mechanical treatment of electron transfer across the interface, differs fundamentally from the differential capacity method (commonly used for determination of the ψ' potential) based on classical considerations. The photoemission measurements have thus validated once more the Gouy–Chapman theory,

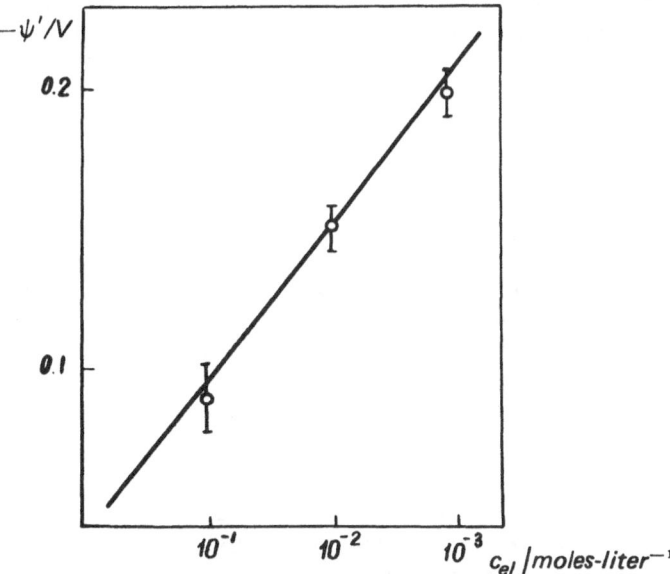

Fig. 6.5. Determination of the ψ' potential (Prishchepa *et al.*, 1975). ψ' potential as a function of electrolyte concentration (HCl). Mercury electrode, potential 1.0 V. Solid line calculated from the Gouy–Chapman theory.

at least for the concentration range 0.01–0.001 M in the absence of specific adsorption.

6.5. Investigation of Specific Adsorption

There are at least two effects of adsorbed ions and molecules on the photocurrent.

1. When specific adsorption of charged particles affects the potential of the outer Helmholtz plane (which will be designated also as ψ'). Table 6.1 indicates qualitatively the effects of specifically adsorbed anions and cations on the photocurrent, if the adsorption-induced changes of the ψ' potential are known. Equations (6.2) and (2.18) quantitatively describe effects of adsorption on photoemission and drift of solvated electrons.
2. When specifically adsorbed ions and molecules block a part of the electrode surface, resulting, in principle, in a change of the quantum yield for photoemission.

Adsorption of halide anions, tetrabutylammonium cations, and organic molecules with long hydrocarbon chains is discussed below. All these particles change, in various ways, the double-layer structure, but a reasonably complete picture of adsorption effects on photoemission can be presented, although the majority of experimental data is only of a semi-quantitative nature.

Anion Adsorption

The changes of the ψ' potential and their effect on photoemission are most pronounced in the case of adsorbed halide ions. The thickness of the adsorbed layer itself is small and can be included in the $\delta < \lambda$ region (where λ is the de Broglie wavelength of the emitted electron). Therefore the process of emission itself should proceed in the same way as from a clean surface. Moreover, the coverage of the surface even with the most-surface active halide, iodide, in its 0.1 M solution does not exceed 0.2–0.3, being lower for other halides (Grahame, 1947). Therefore the primary effect of specific adsorption (e.g., on the hydrogen overpotential) is that of the changed ψ' potential (Frumkin et al., 1952). The same, obviously, pertains to the effect of adsorbed halides on the photocurrent.

The previous discussion of the effect of the ψ' potential on the photocurrent in the absence of specific adsorption remains valid in the present cases. In particular, the effects described above can be observed in dilute solutions. In more concentrated electrolytes, when the diffuse layer thickness is sufficiently small, the photocurrent should virtually be independent of adsorption, as is observed experimentally (cf. Section 4.3, Fig. 4.7).

Photocurrent curves obtained at the mercury electrode in dilute, N_2O–saturated KF, KCl, KBr, and KI solutions are shown in Fig. 6.6. At high negative potentials ($\varphi < -1.1$ V), at which the specific adsorption of halides can be neglected, photocurrents are identical in all solutions. In the specific adsorption region for halides, the photocurrent is lower in KCl, KBr, and KI solutions than in KF. The observed decrease is well correlated with the surface activity of the anion: The photocurrent starts decreasing at more negative potentials in the series Cl^-, Br^-, I^-. At more positive potentials ($\varphi > -0.7$ V) the photocurrent increases somewhat in the case of Br^- and I^-.

The decrease of the photocurrent in the adsorption region from -0.7 to -1.1 V is connected with the effect of the ψ' potential on the emission step. Other effects are absent, since at negative surface charges all emitted electrons remain in the solution owing to the retarding effect of the diffuse-layer field; the photocurrent is then virtually equal to the emission current. Conversely, at potentials approaching the point of zero charge (starting with -0.7 V) in the absence of high acceptor concentrations, a considerable fraction of emitted electrons returns to the electrode. Therefore, anion adsorption, which causes a negative shift of the ψ' potential, primarily affects migration of hydrated electrons in the vicinity of the point of zero charge (Rotenberg and Pleskov, 1970).

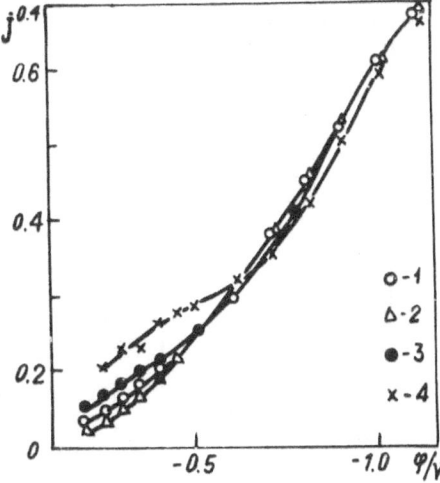

Fig. 6.6. The effect of specific adsorption of halides on the photocurrent (Rotenberg et al., 1973b). Dependence of $j^{0.4}$ on φ at the mercury electrode in 0.1 M solutions of KF (1), KCl (2), KBr (3), and KI (4) saturated with N_2O.

In the presence of cationic acceptors, e.g., hydrogen ions, the negative shift of the ψ' potential results in an increase of H_3O^+ concentration at the interface. Photocurrent vs. potential curves obtained in the presence of iodide ions in acid solutions at an indium electrode are shown in Fig. 6.7. At potentials more negative than $\varphi_{zc} = 0.9$ V, i.e., in the region of negligible iodide adsorption, photocurrents observed in 0.001 M HCl and 0.001 M HCl + 0.001 M KI solutions coincide within the experimental error, as expected. At positive surface charge the ψ' potential changes sign upon adsorption of I^- ions, the migration of hydrated electrons toward the electrode slows down, the interfacial concentration of H_3O^+ ions increases, and the photocurrent increases.

Adsorption of Tetrabutylammonium Cations (TBA)

At surface coverages close to unity, the photocurrent can be affected not only by the changes in the ψ' potential, but also by the screening effect of adsorbed particles.

The effect of adsorption of organic ions on electrochemical kinetics is usually studied using ions of the tetraalkylammonium series, which in a wide potential range form adsorbed monolayers already at relatively low concentrations (cf., e.g., Frumkin, 1961). One of the most effective adsorbates of this type is the tetrabutylammonium cation (TBA). The effects of TBA on electrochemical kinetics consist usually in both the change of ψ' potential and blocking of the surface.

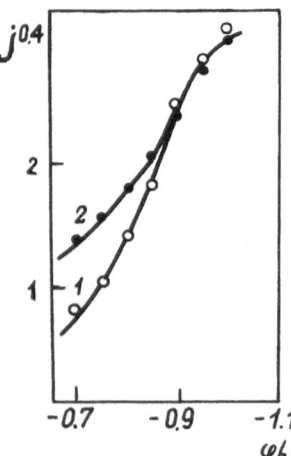

Fig. 6.7.   Effect of specific adsorption of iodide ions on photocurrents (Rotenberg and Pleskov, 1973). Indium electrode, in 0.001 M HCl (1) and 0.001 M HCl + 0.001 M KI (2) solutions.

The separation of the two effects is rather difficult in electrode-kinetic studies carried out in the presence of adsorbed TBA, but is relatively simple in photoemission experiments, as shown below.

The $j^{0.4}$ vs. φ curves measured at mercury in a relatively concentrated (0.1 M) KF solution saturated with N_2O and containing TBA bromide are shown in Fig. 6.8. Upon adsorption of TBA, the photocurrent decreases, the effect being the greater the higher the concentration of TBA. At potentials more positive than -1.1 V the slope of the $j^{0.4}$ vs. φ curve changes, otherwise the type of dependence of photocurrent on potential and the value of the threshold potential remain the same, and in this respect the results in Fig. 6.8 resemble those in Fig. 4.7 for halide ions.

At potentials more negative than -1.1 V, the system behaves in a somewhat peculiar fashion. Desorption of TBA, as measured by the differential capacity method, occurs over a rather narrow potential range (Frumkin and Damaskin, 1959), whereas the increase of the photocurrent to the value observed at a "clean" mercury surface covers a much wider potential range (Rotenberg and Pleskov, 1969). Evidently, the slow increase of the photocurrent reflects not only the desorption of TBA but also the preceding reorientation of adsorbed TBA ions [which appears in the differential capacity curve as an additional small peak localized at potentials

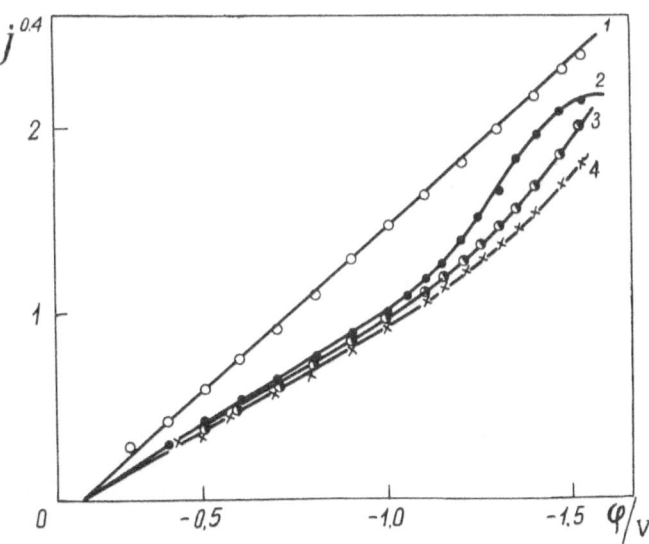

Fig. 6.8. Effect of TBA adsorption on photocurrents (Rotenberg and Pleskov, 1969). Mercury electrode in 0.1 M KF solution saturated with N_2O. TBA concentration (moles-liter^{-1}): (1) 0; (2) 10^{-5}; (3) 10^{-4}; (4) 10^{-3}.

positive to the desorption peak (Damaskin and Nikolaeva-Fedorovich, 1961)].

The effect of TBA adsorption on the photocurrent depends on the concentration of the background electrolyte (Fig. 6.9). In the absence of specific adsorption, the photocurrent passes through a maximum at $c_{el} \simeq$ 0.01 M (curve 1); cf. Fig. 6.1. In the presence of TBA, the j vs. c_{el} curve shows an opposite trend (curve 2); cf. Fig. 6.1. It is obvious that the decrease of photocurrent caused by TBA in dilute solutions is primarily due to the change of sign of the ψ' potential (at positive ψ' values, more electrons return to the surface). A similar effect of the electrolyte concentration on the decrease of the photocurrent in the presence of TBA was also observed at indium electrodes (Rotenberg *et al.*, 1974).

In relatively concentrated electrolytes ($\psi' \simeq 0$), the effect of TBA is small (Fig. 6.9). It is due primarily to the blocking of the electrode surface.

On the whole, the experimental data indicate that the decrease of the photocurrent in the presence of TBA is caused by the change of the ψ' potential rather than by the blocking effect. This is borne out by experimental data concerning effects on the hydrogen overpotential (Krishtalik, 1965). The latter increases considerably upon addition of TBA in the discharge region, where the effect of ψ' potential is most pronounced. Conversely, in a barrierless process, when the ψ' potential does not affect the overpotential, addition of TBA has practically no effect. Thus, both hydrogen ion discharge and photoelectron emission are hindered by adsorbed particles, due mainly to their effect on the ψ' potential.

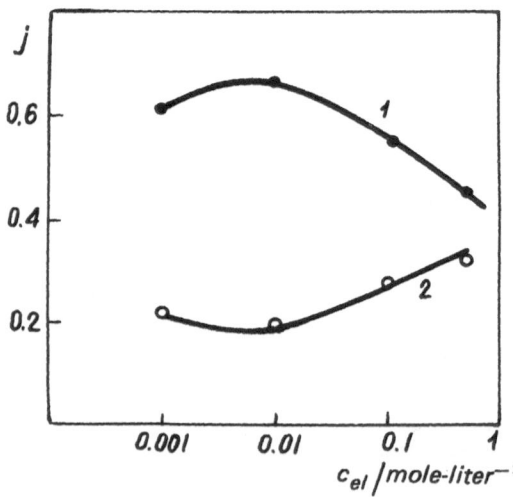

Fig. 6.9. Effect of TBA adsorption on the photocurrent as a function of electrolyte concentration (Rotenberg *et al.*, 1973b). Mercury electrode; KF solutions: (1) saturated with N_2O; (2) with addition of $2 \cdot 10^{-4}$ mole-liter^{-1} TBA; potential 1.0 V.

Adsorption of Organic Molecules with a Long Hydrocarbon Chain

Large organic molecules form at high electrode coverages thick adsorbed layers impenetrable (cf. Section 1.5) with respect to electron emission. Several dissolved organic substances, e.g., alcohols, readily take part in homogeneous reactions with the capture products (H·, OH·); cf. Section 8.3. Therefore, the effects of adsorption of such compounds on photoemission can be studied only after the bulk effects have been excluded. Higher aliphatic alcohols (hexyl, octyl, etc.) and acids which are highly adsorbable and little soluble in water are the most suitable adsorbates for this purpose (Gorodetskaya and Frumkin, 1939; Damaskin *et al.*, 1967).

The experimental $j^{0.4}$ vs. φ curves obtained at a mercury electrode in KCl solution in the presence and absence of decyl alcohol are shown in Fig. 6.10. Adsorption of the alcohol causes a considerable decrease of the photocurrent (20–80 times) in the -0.7 to -1.3 V potential range. Below

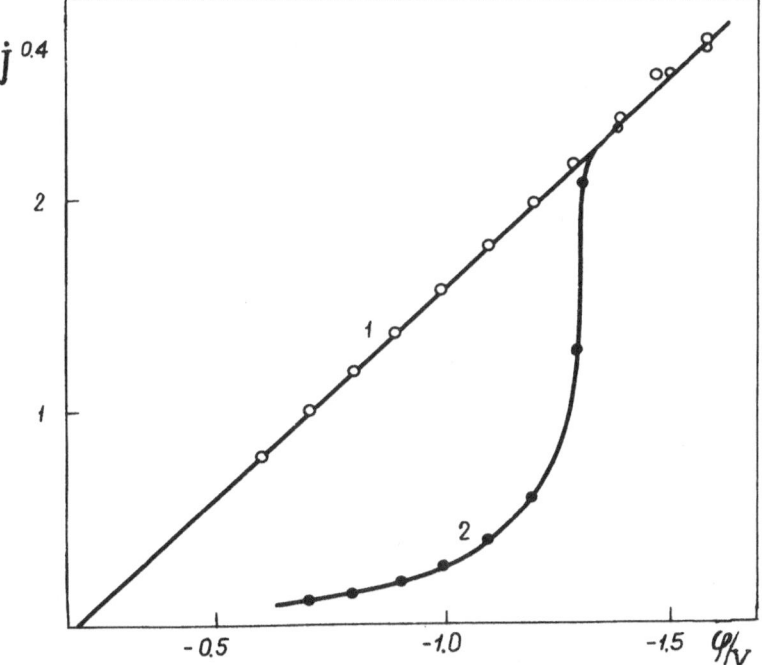

Fig. 6.10. Effect of decyl alcohol adsorption on the photocurrent (Eletskii and Pleskov, 1974). Mercury electrode, 0.1 M KCl solution saturated with N_2O without (1) and with decyl alcohol (2).

−1.4 V, the alcohol fully desorbs from the surface and both curves then coincide. Similar behavior has been observed for other adsorbates of this type (Eletskii and Pleskov, 1974).

The adsorbed molecules are of sufficiently large size to allow results of model considerations to be used (Section 1.5) for the evaluation of the basic parameters of the adsorbed layer. The photoemission current vs. potential curve should not obey the 5/2 power law, but rather an exponential dependence (1.50) should arise as in

$$I = I_{ad} \exp(-b\varphi)$$

where the constants I_{ad} and b are potential independent. It follows (assuming that the I vs. φ and j vs. φ dependences differ little) that the experimental data should yield a straight line in a plot of $\ln(j/I_{ad})$ vs. φ. This is indeed the case. The plot of $\ln[j(E_m^0)^2/I_{ad}]$ vs. φ, where $E_m^0 = \hbar\omega - \hbar\omega_0(0)$ and $\hbar\omega_0(0)$ refers to the potential of zero charge in the absence of specific adsorption, is shown in Fig. 6.11. From the parameters of the resulting straight line, the thickness of the adsorbed layer d_{ad} and the height of the potential barrier U_{ad} can be calculated as functions of the dielectric constant ε_{ad} of the adsorbed layer.

Evidently, the thickness d_{ad} depends little on the ε_{ad} value. In the case of cetyl alcohol, variation of the dielectric constant in the $1 < \varepsilon_{ad} < \infty$ range results in d_{ad} values varying from 17 to 20 Å. The d_{ad} value also depends little on the detailed shape of the potential drop $V(x)$ and is determined only by its integral characteristics. Therefore, the choice of the potential model is not essential for determination of d_{ad}. On the other hand, the U_{ad} value can be considered as a certain averaged characteristic of the adsorbed layer (cf. Section 1.5).

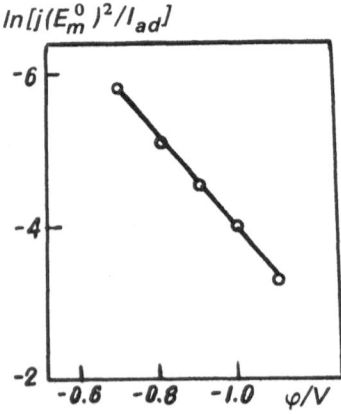

Fig. 6.11. Graphical treatment of the j vs. φ curve in the presence of decyl alcohol (Fig. 6.10) in terms of Eq. (1.50) (Eletskii and Pleskov, 1974).

Table 6.3. Adsorption Characteristics of Some Alcohols at a Mercury Electrode

Alcohol	d_{ad} (Å)	ε_{ad}	U_{ad} (eV)
Hexyl	9.3	4.1	0.7
Octyl	11.7	3.2	0.6
Decyl	12.7	—	0.7
Cetyl	18.0	2.0	0.8

Combination of the two methods, photoemission and differential capacity, allows the adsorbed layer thickness and dielectric constant to be separately established. Results of such estimates are shown in Table 6.3 (Eletskii and Pleskov, 1974).

The monolayer thickness is close to that calculated (based on the lengths of C–C and C–O bonds) from the respective lengths of the alcohol molecules. Thus, at high coverages ($\theta = 1$) hydrocarbon chains are oriented perpendicularly to the electrode surface. The monotonic decrease of the dielectric constant with increasing length of the hydrocarbon chain can be explained by the decreasing amount of water in the increasingly hydrophobic adsorbed layers.

In concluding this section, we should mention the strong effect of adsorbed hydrogen on the photoemission current at a platinum electrode (Ansone *et al.*, 1976). The experimental data, however, have not yet been conclusively interpreted.

6.6. Evaluation of the Thickness of the Compact Layer

It has been assumed until now that the parameters of the compact part of the double layer have a negligible effect on the overall photoemission process. Meanwhile, deviations from the 5/2 power law observed in certain cases, which cannot be explained in terms of other effects, can be ascribed to a finite thickness of the compact layer.

Rotenberg *et al.* (1973a) estimated the thickness of the Helmholtz layer at a mercury electrode in 0.1 *M* NaF solution to be $\simeq 2$ Å. This is much less than the sum of the radius of the Na^+ ion and the diameter of a water molecule, i.e., of the total thickness in the model of the Helmholtz layer which assumes the existence of an adsorbed monolayer of water. However, in terms of the estimate mentioned above, the result does not represent the total distance between the metal surface and the outer Helmholtz plane, but only that part over which the potential drops sharply. According to the calculations of Krishtalik (1972) carried out for the model consisting of metal–monolayer of water–ionic plate of the double layer, the main potential drop ($\simeq 90\%$) is confined to the monolayer of water (with a lower dielectric constant). Further potential changes, up to the outer Helmholtz plane, are much slower. The $\simeq 2$ Å thickness of the compact layer is in agreement with Krishtalik's calculations.

It should be stressed that the results described in this chapter indicate that the properties of the double layer and also specific adsorption of ions and molecules can be studied by means of the photoemission method. The agreement of the experimental results obtained in this way with those provided by traditional methods is of fundamental importance and confirms the validity of the present concepts concerning the structure of the metal–electrolyte interface.

Photoemission As a Method of Investigation of Electrochemical Kinetics: Processes Involving Atomic Hydrogen

Homogeneous and heterogeneous processes involving free radicals play an important role among reactions of solvated electrons initiated by photo-electron emission. The kinetics of these processes affect the magnitude of the photocurrent, thus allowing photoemission to be used as a method for their quantitative investigation.

This chapter is concerned with a detailed discussion of electrochemical reactions involving atomic hydrogen. In the next chapter, applications of the photoemission method for studies of homogeneous chemical reactions will be considered.

7.1. Formation of Atomic Hydrogen in Electrochemical Reactions. Phenomenology and Empirical Equations

Atomic hydrogen and its electrochemical properties have, for a long time, evoked lively interest among electrochemists. This is understandable as it is an intermediate in one of the most thoroughly studied reactions — electrochemical hydrogen evolution. Photoemission in acid solutions also results in the formation of hydrogen *atoms* as the products of capture by ions of hydrated electrons (only aqueous solutions will be considered below):

$$e_{aq}^- + H_3O^+ \rightarrow H^{\cdot} + H_2O \tag{7.A}$$

Free hydrogen atoms cannot exist in the vicinity of the electrode: They either recombine with each other, or enter electrochemical reactions at the metal surface. The first path — recombination of atomic hydrogen — has a low probability, being a second-order reaction proceeding at very slow rates at the practically attainable concentrations of hydrogen atoms. Therefore, virtually all hydrogen atoms produced in the solution return to the electrode and are electrochemically removed.

At metals weakly adsorbing hydrogen, the main path is of the electro-chemical desorption type (cf. Section 2.3)

$$H^{\cdot} + HS + e^{-}(M) \rightarrow H_2 + S^{-} \qquad (7.B)$$

where HS is a proton source which, in aqueous solution, may be hydrogen ions or water molecules. At more positive potentials, a parallel reaction of ionization

$$H^{\cdot} + H_2O \rightarrow H_3O^{+} + e^{-}(M) \qquad (7.C)$$

can take place, resulting, as in all anodic reactions, in a decrease of the photocurrent ("hydrogen drop"). The reaction was first found to proceed in photoemission measurements by Barker et al. (1966) and was studied in more detail by Pleskov et al. (1974) and Rotenberg et al. (1970a, 1974, 1975b) at mercury and solid electrodes.

Apart from the reactions mentioned above, another possibility for hydrogen atom removal consists in their interaction in solution with hydrogen ions to form the radical ion H_2^{+}:

$$H^{\cdot} + H_3O^{+} \rightarrow H_2^{+} + H_2O \qquad (7.D)$$

which is relatively easily reduced at the electrode. However, the rate constant of this reaction ($k_r = 10^4$ mole^{-1}-liter-sec^{-1}), established by radiation-chemical measurements (Dolin and Ershler, 1962), is too low to play any significant part in the overall process. It should be mentioned that the formation of the H_2^{+} ion intermediate in the electrochemical hydrogen evolution was considered earlier by Horiuti (1958) and Matsuda and Horiuti (1958). Their hypothesis, however, found no experimental confirmation on metals with a high hydrogen overpotential such as mercury, lead, or bismuth. Recombination of hydrogen atoms at the electrode surface can be neglected in the case of the latter metals.

It should be mentioned that the simplicity of the method and the easy quantitative control of the rate of supply of atomic hydrogen makes photo-emission a more convenient source of the latter than the methods previously used (Frumkin, 1957; Levina and Kalish, 1956; Bagotskaya and Oshe, 1959). The photoemission method is closest to that applied by Levina and Kalish (1956) who produced atomic hydrogen by discharge in the gas phase with subsequent diffusion through a thin solution layer to the electrode surface.

Taking into account reactions (7.B) and (7.C), the measured photo-current j is given by

$$j = j_0 + \vec{j}_H - \overleftarrow{j}_H \qquad (7.1)$$

where $j_0 = I - I_e$ is the emission current (after subtraction of the reverse current of electrons which escape capture by H_3O^+ ions); $\vec{j}_H$ is the cathodic current of reaction (7.B) and $\overleftarrow{j}_H$ is the anodic current of reaction (7.C). As has been already mentioned, two mechanisms of reactions (7.B) and (7.C) are conceivable: (a) through the adsorbed intermediate H_{ad}, and (b) directly from solution, without passing through the adsorbed state. [Reaction (7.B) will be called further electrochemical desorption regardless of the state, adsorbed or dissolved, of the atomic hydrogen.] When reaction occurs along both paths simultaneously, the steady-state equations can be written as

$$j_0 = (k_{ad} + \vec{k}_1 + \overleftarrow{k}_1)c_H(0) \qquad (7.2a)$$
$$k_{ad}c_H(0) = (\vec{k}_2 + \overleftarrow{k}_2)\theta \qquad (7.2b)$$
$$j = j_0 + (\vec{k}_1 - \overleftarrow{k}_1)c_H(0) + (\vec{k}_2 - \overleftarrow{k}_2)\theta \qquad (7.2c)$$

where $c_H(0)$ is the hydrogen atom concentration in the vicinity of the electrode; θ is the surface coverage with adsorbed hydrogen, k_{ad} is the rate constant of adsorption expressed (in a similar way to that for other kinetic constants) in electrical units, k_i are the rate constants of electrochemical reactions; $i = 1, 2$. The indices 1 and 2 on the quantities k, α, and β refer in this chapter to reactions of hydrogen atoms from the solution (1) and from the adsorbed state (2). Also, the rate constants $\vec{k}_1$ and $\overleftarrow{k}_1$ pertain to the cathodic and anodic removal, respectively, of hydrogen atoms directly from solution, while $\vec{k}_2$ and $\overleftarrow{k}_2$ refer to adsorbed hydrogen. The potential dependence of rate constants can be expressed (Frumkin *et al.*, 1952; Delahay, 1965) by

$$\vec{k}_i = \vec{k}_{i0} \exp(-\alpha_i F\varphi/RT), \qquad \overleftarrow{k}_i = \overleftarrow{k}_{i0} \exp(\beta_i F\varphi/RT)$$

where $\vec{k}_i$ and $\overleftarrow{k}_{i0}$ are the rate constants at $\varphi = 0$. The rate constant of adsorption k_{ad} is, in the first approximation, potential independent, F is the Faraday number, R is the gas constant, and α_i and β_i are the transfer coefficients.

When reaction (7.B) proceeds with the simultaneous participation of water molecules and hydrogen ions, the rate constants $\vec{k}_1$ and $\vec{k}_2$ can be formally divided into two terms $\vec{k}_i = k'_i + Xk''_i$ where k'_i and k''_i are the rate constants of hydrogen removal by water and by hydrogen ions, respectively, and X is the mole fraction of H_3O^+ ions in the solution. (The possibility of the electrochemical removal of adsorbed hydrogen by water molecules in acid solutions was first indicated by Krishtalik, 1968.)

Elimination of θ and $c_H(0)$ from Eqs. (7.2) results in

$$j = 2j_0\left[\frac{\vec{k}_1}{k_{ad} + \vec{k}_1 + \overleftarrow{k}_1} + \frac{\vec{k}_2 k_{ad}}{(k_{ad} + \vec{k}_1 + \overleftarrow{k}_1)(\vec{k}_2 + \overleftarrow{k}_2)}\right] \qquad (7.3)$$

In the limiting case, when the oxidation reaction can be neglected ($\overleftarrow{k}_1 = \overleftarrow{k}_2 = 0$), we obtain $j = 2j_0$; i.e., the usual dependence of photocurrent on potential is observed. We can mention here, in advance, that the same potential dependence of the photocurrent (i.e., the 5/2 power law in concentrated acids) obtains in the case of potential independent rate constants in Eq. (7.3), for example, in nonactivated processes (see below). In general, the rate constants in Eq. (7.3) are potential dependent and the j vs. φ dependence is more complex. For the sake of comparison of the theory with experiment, we shall consider now some limiting cases:

(a) Reactions (7.B) and (7.C) proceed without adsorption ($k_{ad} = 0$). Then

$$j = 2j_0 \frac{\vec{k}_1}{\vec{k}_1 + \overleftarrow{k}_1}$$

Taking into account the potential dependence of rate constants $\vec{k}_1$ and $\overleftarrow{k}_1$, we obtain

$$\varphi = \frac{2.3RT}{(\alpha_1 + \beta_1)F} \left(\log \frac{2j_0 - j}{j} + \log \frac{\vec{k}'_{10} + \vec{k}''_{10}X}{\overleftarrow{k}_{10}} \right) \tag{7.4}$$

Equation (7.4), as well as (7.5) and (7.7), was derived assuming the transfer coefficient α_i in reaction (7.B) to be independent of the nature of the HS reactant. In the general case (see below), the two mechanisms of electrochemical desorption should be reflected in two different values of the transfer coefficient α.

(b) Only adsorbed hydrogen participates in both reactions ($\vec{k}_1 = \overleftarrow{k}_1 = 0$). Then

$$\varphi = \frac{2.3RT}{(\alpha_2 + \beta_2)F} \left(\log \frac{2j_0 - j}{j} + \log \frac{\vec{k}'_{20} + \vec{k}''_{20}X}{\overleftarrow{k}_{20}} \right) \tag{7.5}$$

(c) Reaction (7.B) involves the absorption step, as opposed to reaction (7.C) ($\vec{k}_1 = \overleftarrow{k}_2 = 0$). Then

$$\varphi = \frac{2.3RT}{\beta_1 F} \left(\log \frac{2j_0 - j}{j} + \log \frac{k_{ad}}{\overleftarrow{k}_{10}} \right) \tag{7.6}$$

(d) Reaction (7.B) proceeds without adsorption and (7.C) involves adsorbed hydrogen ($\vec{k}_2 = \overleftarrow{k}_2 = 0$). Then

$$\varphi = \frac{2.3RT}{\alpha_1 F} \left(\log \frac{2j_0 - j}{j} + \log \frac{\vec{k}'_{10} + \vec{k}''_{10}X}{k_{ad}} \right) \tag{7.7}$$

Equations (7.4)–(7.7) represent all mechanisms of atomic hydrogen removal encountered in reality. Their form resembles that of the kinetic equations for slow discharge, from which they differ in that the discharge current is replaced by a dimensionless parameter $(2j_0 - j)/j$, equal to the ratio of hydrogen oxidation current ($\overleftarrow{j}_H$) to its electroreduction current ($\vec{j}_H$).

This is easy to demonstrate using the relation $j_0 = |\vec{j}_H| + |\overleftarrow{j}_H|$ which follows from the law of charge conservation. It follows from Eqs. (7.4)–(7.7) that $\log[(2j_0 - j)/j]$ should be a linear function of φ with a slope depending on the transfer coefficients of the respective reactions.

The above equations can be easily modified to account, if necessary, for the double-layer effects. This requires introduction of the ψ' potential into the expressions for $\vec{k}_1$ and $\overleftarrow{k}_1$ and replacement of the bulk concentration of hydrogen ions by a ψ'-dependent expression. This will be done in Section 7.5 in the description of experimental results.

7.2. Basic Experimental Data Concerning Atomic Hydrogen Reactions

Potential Dependence

Quantitative experimental photoemission studies of the electrochemical kinetics of atomic hydrogen reactions have been carried out on two metals only, mercury and bismuth. Qualitative data were also obtained for antimony. At other metals (lead, cadmium, indium) the "hydrogen drop," indicating the starting point of the atomic hydrogen oxidation, was not observed. Evidently, the potential region of this process, if it occurs at all, is more positive than the stationary potentials of these metals. The $j^{0.4}$ vs. φ curves obtained for the photocurrent on mercury and bismuth in solutions containing hydrogen ions as acceptors are shown in Figs. 7.1 and 7.2. The plots deviate from linearity, in the absence of surface active substances, at potentials

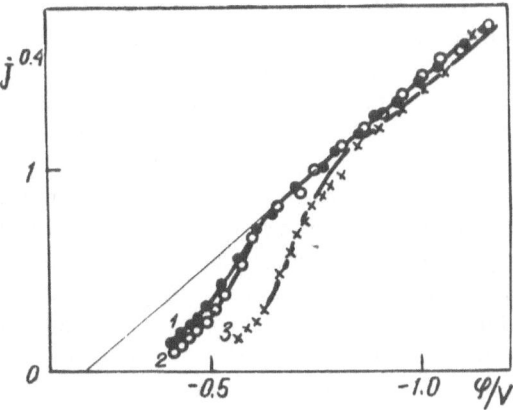

Fig. 7.1. Reactions of atomic hydrogen at a mercury electrode. The effect of halide ions (Rotenberg *et al.*, 1970a). Solutions: (1) 0.09 *M* KCl; (2) 0.09 *M* KBr; (3) 0.09 *M* KI, with addition of 0.01 *M* HCl.

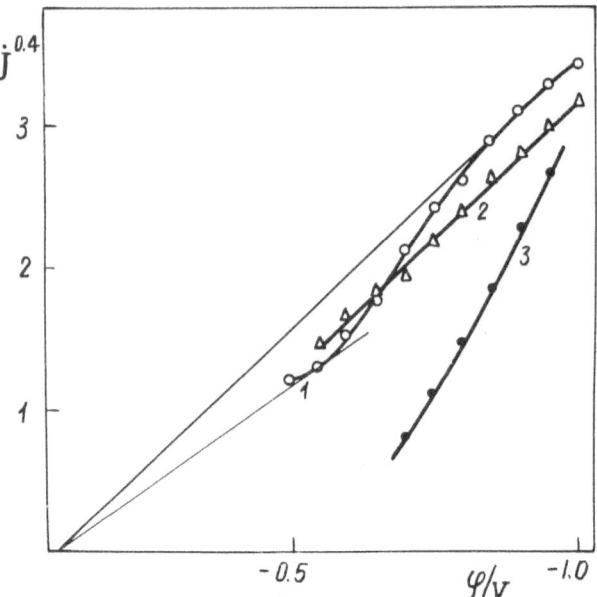

Fig. 7.2. Reactions of atomic hydrogen at a bismuth electrode (Rotenberg *et al.*, 1974). (1) 0.1 *M* HCl + 0.4 *M* KCl; (2) 0.1 *M* HCl + 0.4 *M* KCl + 2×10^{-4} *M* TBA; (3) 0.1 *M* HCl + 0.2 *M* KI.

above -0.6 to -0.7 V on mercury, and above -0.8 V on bismuth. On antimony the photocurrent decreases around -0.1 V; it is impossible to establish accurately the region where these deviations from the linear dependence occur owing to the cathodic discharge of hydrogen ions which commences in the same region (Rotenberg *et al.*, 1975b).

The experimental verification of Eqs. (7.4)–(7.7) requires not only knowledge of the dependence of j on φ but also of the $2j_0$ current in the absence of the oxidation current of atomic hydrogen. The latter can be found only by extrapolation of the linear section of the dependence of $j^{0.4}$ on φ from the region of negative potentials into the nonlinear region, or by measurements of photocurrents in the presence of other acceptors, e.g., N_2O, which are not electro-oxidized after electron capture. The latter acceptor forms an $OH^{\cdot}$ radical upon electron capture, which becomes reduced at a high rate at the electrode over the entire potential range. Comparison of photocurrents obtained in solutions of hydrogen ions and N_2O makes possible a calculation of the $(2j_0 - j)/j$ value and consequently a construction of its experimental potential dependence. Such plots are shown in Figs. 7.3 and 7.4 for mercury and bismuth, respectively. The plot obtained for mercury

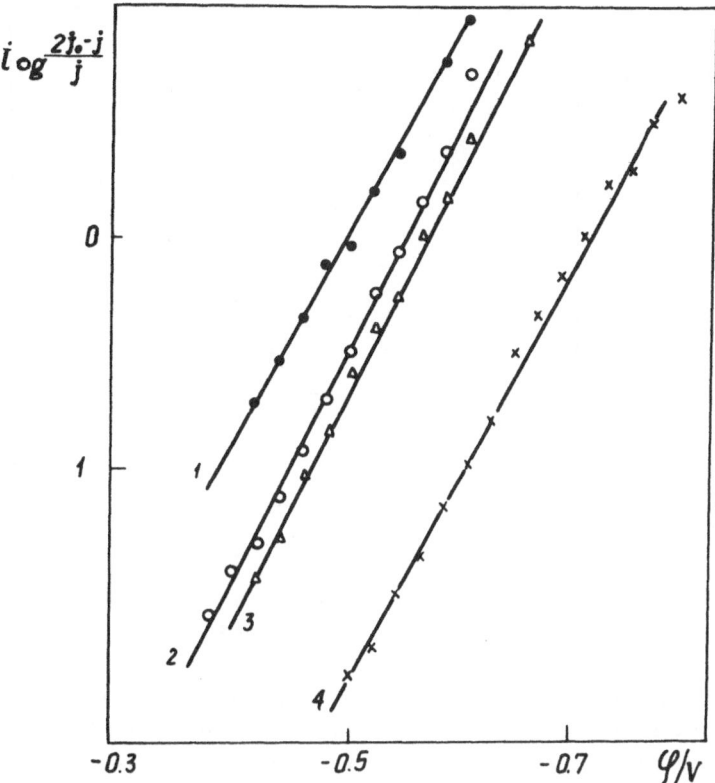

Fig. 7.3. Plot of $\log[(2j_0 - j)/j]$ vs. φ obtained at a mercury electrode (Rotenberg *et al.*, 1970a). Solutions: (1) 0.05 M K$_2$SO$_4$; (2) 0.1 M KCl; (3) 0.1 M KBr; (4) 0.1 M KI, with addition of 0.005 M H$_2$SO$_4$.

is linear over the entire potential range studied, indicating that the kinetics of removal of atomic hydrogen obey Eqs. (7.4)–(7.7). The same plot obtained at a bismuth electrode is linear over the range -0.8 to -0.6 V; above -0.6 V the plot becomes virtually potential independent. The slope of the straight lines are 100–120 mV for mercury and about 140–150 mV for the bismuth electrode (below -0.6 V).

Effects of Specific Adsorption

The kinetics of the electrochemical removal of atomic hydrogen depend to a great extent on the structure of the double layer and on solution composition. It can be seen from Figs. 7.1 and 7.2 that the specific adsorption

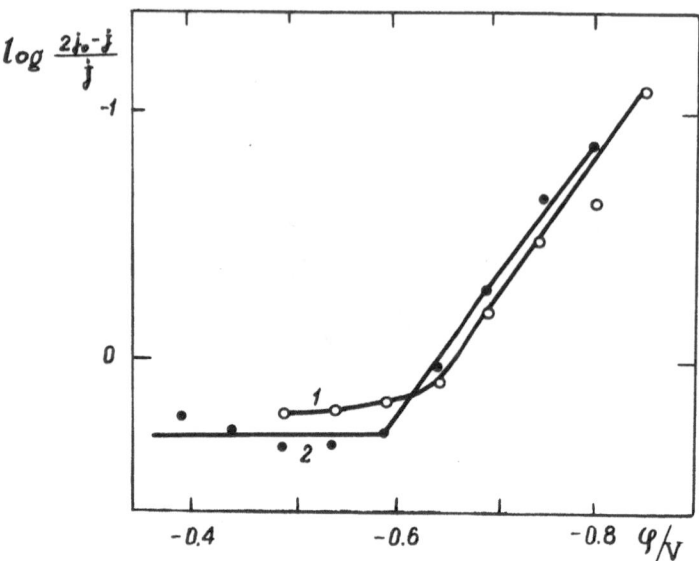

Fig. 7.4. Plot of $\log[(2j_0 - j)/j]$ vs. φ obtained at a bismuth electrode (Rotenberg *et al.*, 1974). (1) 0.5 *M* KBr + 0.003 *M* HCl; (2) 0.5 *M* KCl + 0.06 *M* HCl.

of halide ions results in an increase of the ratio $\overleftarrow{j}_H/\overrightarrow{j}_H$ at the given potential ("the hydrogen drop" is shifted toward more negative potentials). Adsorption of the tetrabutylammonium ion (TBA) causes a decrease of the photocurrent in the whole potential range, and the "hydrogen drop" practically disappears. (It should be recalled that the decrease of photocurrent upon TBA adsorption, which occurs both in acid and N_2O solutions, is due to the effect of adsorption on the act of emission itself and on the migration of hydrated electrons.)

The pH Effect

If one of the parallel steps of removal of atomic hydrogen is reaction (7.B), the kinetics of the overall process should, according to Eqs. (7.4), (7.5), and (7.7), depend in principle, on the concentration of hydrogen ions in the solution. It is convenient to denote the potential at which cathodic and anodic currents are equal, i.e., $(2j_0 - j)/j = 1$, by φ^*. The potential φ^* can be found from the intersection point of the experimental $j^{0.4}$ vs. φ curve with a straight line drawn through the point $\varphi = \varphi_0$ with a slope $2^{0.4}$ times lower than that of the experimental line in the cathodic potential region.

The dependence of φ^* on hydrogen ion concentration, $c_{H_3O^+}$, for 0.001 to 1 M solutions is shown in Figs. 7.5 and 7.6. The potential φ^* increases monotonically on mercury with decreasing pH, the sharpest change occurring in the high concentration range. Below 0.01 M, φ^* depends very little on $c_{H_3O^+}$. On bismuth the pH effect is much less pronounced. It can be seen from Fig. 7.6 that within experimental error the value of φ^* remains constant with pH over the concentration range 10^{-3} to 1 M.

Qualitatively, the dependence of φ^* on pH at mercury indicates that both hydrogen ions and water molecules participate in the electrochemical desorption step. From Eqs. (7.4), (7.5), or (7.7) and Figs. 7.5 and 7.6, the ratio of the rate constants of the cathodic removal of hydrogen atoms by hydrogen ions and water molecules can easily be determined. For example, using Eq. (7.5) in the form

$$\varphi^* = b \log\left(\frac{\vec{k}'_{20}}{\overleftarrow{k}_{20}} + \frac{\vec{k}''_{20}}{\overleftarrow{k}_{20}} X\right) \tag{7.8}$$

we can find from the experimental dependence of φ^* on X the ratios $\vec{k}'_{20}/\overleftarrow{k}_{20}$ and $\vec{k}''_{20}/\overleftarrow{k}_{20}$ and therefore $\vec{k}''_{20}/\vec{k}'_{20}$. The quantity b is equal to the slope of the φ vs. $\log[(2j_0 - j)/j]$ line (cf. Figs. 7.3 and 7.4). The ratio $\vec{k}''_{20}/\vec{k}'_{20}$ amounts to 1600 on mercury and 55 on bismuth. Thus, in 1 M acid solutions ($X = 1/55$) the removal of atomic hydrogen by hydrogen ions is the main path of the electrochemical desorption on mercury. The value of the slope of the φ^* vs. $c_{H_3O^+}$ line in the high concentration region (~ 100 mV) confirms this conclusion. Comparison of Eqs. (7.4) and (7.8) shows that mechanism (7.B) results in the same slopes of the φ^* vs. $\log(c_{H_3O^+})$ and φ vs. $\log[(2j_0 - j)/j]$ lines. Only for hydrogen ion concentrations below 0.01 M is the ratio $X\vec{k}''_{20}/\vec{k}'_{20}$ much less than unity so that hydrogen ions then virtually take no part in the overall process.

Hydrogen ions participate to a lesser extent in the electrochemical

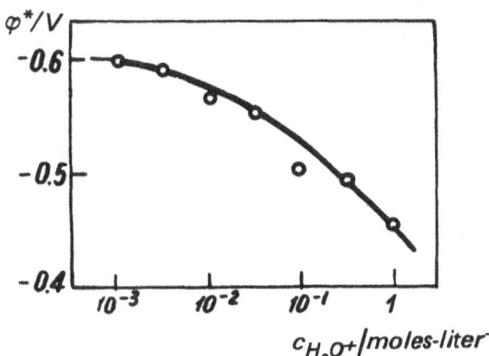

Fig. 7.5. Dependence of φ^* on hydrogen ion concentration. Mercury electrode (Rotenberg et al., 1970a).

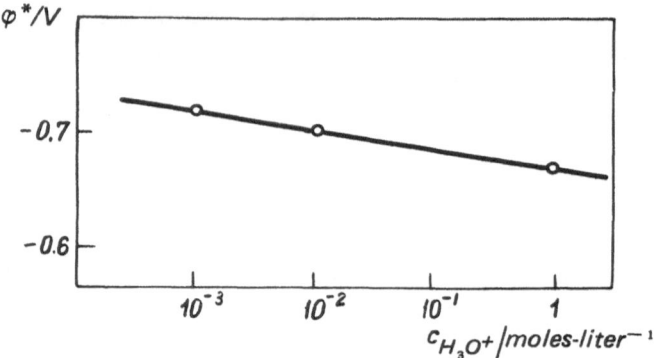

Fig. 7.6. Dependence of φ^* on hydrogen ion concentration. Bismuth electrode (Rotenberg *et al.*, 1974).

desorption step on bismuth electrodes. In 1 *M* HCl solution, hydrogen ions and water molecules are approximately equally active in reaction (7.B). At lower concentrations, electrochemical desorption proceeds predominantly through water molecules.

7.3. The Role of the Adsorption Stage

The kinetic equations in Section 7.1 were derived assuming two main paths for removal of atomic hydrogen from the electrode surface. The first involves direct electron transfer between the hydrogen atom in solution and the metal. The second mechanism includes a step in which adsorption of H atoms precedes the electrochemical reactions. Both types of electrochemical reactions, ionization of atomic hydrogen and its electrochemical desorption, have considerable reaction energies, which can be evaluated by considering suitable cycles.

For the direct removal of dissolved hydrogen atoms (7.C) the reaction energy at equilibrium is equal to

$$q_{\text{ion}} = \Delta H - q \tag{7.7a}$$

where ΔH is the enthalpy of formation of hydrogen atoms from $\frac{1}{2}$ mole of H_2 at 25°C, viz. 52 kcal-mole^{-1} (M. Kh. Karapetyants and M. L. Karapetyants, 1968) and q is the latent heat of the electrode process

$$H_3O^+ + e^-(M) \rightarrow \tfrac{1}{2}H_2 + H_2O \tag{7.E}$$

at equilibrium.

For electrochemical desorption in which H_3O^+ is the source of protons, the heat of desorption q_{des} is given by

$$q_{des} = \Delta H + q \qquad (7.7b)$$

It was believed for a long time that q cannot be directly obtained from experiment (cf., e.g., Temkin, 1953). In recent years, however (Krishtalik, 1969), the value of q was successfully determined from kinetic data on hydrogen overpotential. The method, suggested by Krishtalik, is based on the principle that the ideal activation energies of ordinary and barrierless hydrogen discharge become equal at a certain overpotential η', and then

$$q = -[2(A_e^{(or)} - A_e^{(bl)}) + \eta'F] \qquad (7.8a)$$

where $A_e^{(or)}$ and $A_e^{(bl)}$ are the real activation energies of H_3O^+ discharge by the ordinary and barrierless mechanism, at $\eta = 0$, respectively. The value of η' is obtained from the intersection point of the two Tafel lines for the ordinary and barrierless discharge mechanisms.

The value of $A_e^{(bl)}$ on mercury amounts to 22.9 kcal-mole^{-1} (Krishtalik, 1965). On the basic of the experimental data of Jofa and Mikulin (1944) for the temperature dependence of hydrogen overpotential on mercury in $0.25\ N\ H_2SO_4 + 1\ N\ Na_2SO_4$ solution, Temkin calculated the value of $A_e^{(or)}$ as 21.7 kcal-mole^{-1}. Extrapolation of the overpotential curve obtained in this solution to the intersection with the barrierless discharge line results in $\eta' = 177$ mV (Krishtalik, 1969). From Eq. (7.8a) the value of q is 1.7 kcal-mole^{-1}; then, taking into account the pH of the solution studied, at the potential of the normal hydrogen electrode $q_0 = -3.6$ kcal-mole^{-1}.

From Eqs. (7.7a) and (7.7b) the values of q_{ion} and q_{des} of both parallel reactions considered can be easily found. They amount to 55.6 kcal-mole^{-1} for the ionization, and 48.4 kcal-mole^{-1} for the electrochemical desorption reaction.

When both reactions (7.B) and (7.C) include adsorption steps, the respective values of q_{ion} and q_{des} are diminished by the energy of the bond between atomic hydrogen and the metal (its heat of adsorption). The latter quantity for mercury is 29 kcal-mole^{-1} (Krishtalik, 1965). Consequently, in this case $q_{ion} = 26.6$ and $q_{des} = 19.4$ kcal-mole^{-1}.

Thus, reactions (7.B) and (7.C) proceeding without adsorption are extremely exothermic and their activation energies should be zero. These processes are thus undoubtedly nonactivated ($\alpha_1 = \beta_1 = 0$) and their rate constants are potential independent.

This conclusion can also be reached on the basis of experimental data concerning hydrogen overpotential on mercury (Krishtalik, 1965, 1968). In particular, the electrochemical desorption at the equilibrium potential was convincingly demonstrated to be nonactivated. Under these conditions, hydrogen resulting from H_3O^+ ion discharge was found to be in the adsorbed state, $H_{ad}^.$; thus, its energy is lower than that of the free hydrogen atom in solution by 29 kcal-mole^{-1}. Therefore, there is no doubt that reaction (7.B) involving a free H atom is nonactivated over the entire potential range applied in photocurrent measurements.

Ionization of adsorbed atomic hydrogen is the reverse reaction to H_3O^+ discharge. If the ionization of $H_{ad}^.$ proceeds according to the ordinary mechanism (i.e., with a finite activation energy), the discharge of H_3O^+ ions must also be activated. The nonactivated mechanism of $H_{ad}^.$ ionization corresponds to the barrierless discharge of H_3O^+ ions.

The electrode potential at which ordinary discharge changes over into a barrierless one does not depend on the hydrogen ion concentration and is solely determined by the structure of the double layer (value of the ψ' potential) and the adsorption energy of atomic hydrogen.

The transition from ordinary to barrierless discharge occurs, as was already mentioned, on mercury in $H_2SO_4 + 1\ N\ Na_2SO_4$ solution at $\eta' = 177\ mV$ ($-0.5\ V$ against the saturated calomel electrode). Taking into account the energy of adsorption of atomic hydrogen on mercury (29 kcal-mole^{-1}, $\equiv 1.26\ eV$), the conclusion must be reached that ionization of dissolved atomic hydrogen should be nonactivated down to $-1.76\ V$ (i.e., over the entire potential range applied in photocurrent measurements).

The nonactivated mechanism of reactions (7.B) and (7.C) proceeding with dissolved hydrogen applies to all metals, since for this path the initial and final states of reactants are independent of the adsorption energy.

Thus, if both coupled reactions proceed in solution, the ratio of their rates should be potential independent, a conclusion which contradicts the experimental data presented in Figs. 7.4 and 7.3. Therefore the mechanism of electrochemical removal of atomic hydrogen which excludes the adsorption step seems to be unrealistic. Two other mechanisms, assuming one of the coupled reactions to proceed in solution and the other to include adsorption [Eqs. (7.6) and (7.7)] are also contradicted by experiment. In either case, the transfer coefficients which appear in Eqs. (7.7) and (7.6) should be zero; this is not observed experimentally.

Thus, the conclusion must be reached that the mechanism of removal of atomic hydrogen includes the steps

$$H^.(\text{solution}) \rightarrow H_{ad}$$
$$H_{ad} + HS + e-(M) \rightarrow H_2 + S^- \qquad \qquad (7.B)$$
$$H_{ad} + H_2O \rightarrow H_3O^+ + e^-(M) \qquad \qquad (7.C)$$

described by the kinetic equation (7.5).

The existence of an intermediate state of adsorption of atomic hydrogen was also deduced by Barker (1971) who studied photocurrents in acid solutions containing ethanol.

7.4. The Mechanism of Atomic Hydrogen Reactions on Mercury Electrodes

One of the characteristic features of the overall process of removal of atomic hydrogen is the potential dependence of the ratio of oxidation current to electrochemical desorption current. It can be seen from Fig. 7.3 that for mercury in various solutions the slope of the φ vs. $\log[(2j_0 - j)/j]$ line, in other words, of the φ vs. $\log(\overleftarrow{j}/\overrightarrow{j})$ line, is close to 100–120 mV. It follows from Eq. (7.5) that the sum of the transfer coefficients for reactions (7.B) and (7.C) is close to $\frac{1}{2}$.

It should be mentioned here that, thermodynamically, the sum of the transfer coefficients for the forward and back reactions must be unity (or, more strictly, equal to the number of electrons transferred in the elementary act). Since electrochemical desorption of atomic hydrogen is not a reaction which is the reverse of ionization, the above rule is obviously not applicable. Therefore, the sum of the transfer coefficients, in general, is by no means equal to unity.

The experimentally evaluated sum $\alpha_2 + \beta_2$ can be interpreted in two ways, either assuming the transfer coefficients of both reactions to be lower than $\frac{1}{2}$ (e.g., 0.2 and 0.3), or assuming one of the reactions to be nonactivated, i.e., with the transfer coefficient equal to zero and the second reaction having a transfer coefficient equal to $\frac{1}{2}$.

It is of interest to inquire how the two possibilities might be distinguished electrochemically. Although the transfer coefficient α_2 for electrochemical desorption of hydrogen on mercury is, in general, unknown, the transfer coefficient of the ionization reaction can be easily determined if that of the reverse reaction, hydrogen ion discharge, α, is known. The latter reaction has been investigated on mercury in detail by many authors. The value of $\alpha = \frac{1}{2}$ has been well established for the more negative potential region, and at less negative potentials, when the discharge becomes barrierless, $\alpha = 1$. Since $\alpha + \beta_2 = 1$, the transfer coefficient β_2 can be either zero or $\frac{1}{2}$. Since the sum $\alpha_2 + \beta_2 = \frac{1}{2}$, α_2 can also be either zero or $\frac{1}{2}$. In other words, one of the two reactions, electrochemical desorption or ionization, is nonactivated.

The experimental data for mercury electrodes described in previous sections cannot resolve unambiguously the question which of the two reactions is nonactivated; only some indirect arguments for the actual mechanism of the process can be invoked. Before discussing them, let us consider the basic diagnostic criterion, namely the character of the dependence of $\log[(2j_0 - j)/j]$ (or $\log(\overleftarrow{j_H}/\overrightarrow{j_H})$ on potential. Three main types of behavior can be envisaged for the coupled reactions of ionization and electrochemical desorption; they are shown schematically in Fig. 7.7. With increasing cathodic overpotential, the rate of electrochemical desorption increases and the reaction becomes nonactivated ($\alpha_2 = 0$) at high negative

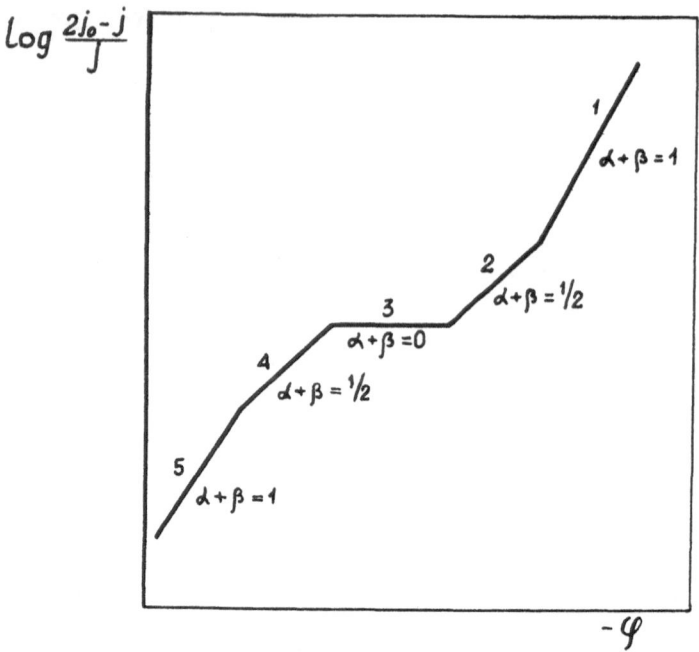

Fig. 7.7.   Potential dependence of $\log[(2j_0 - j)/j]$ for various mechanisms of atomic hydrogen reactions.

potentials. Simultaneously the rate of ionization decreases with increasing negative potential, and if conditions are reached such that the activation energy of ionization is equal to the reaction energy, the process becomes barrierless ($\beta_2 = 1$). Therefore the limiting slope of the $\log[(2j_0 - j)/j]$ vs. φ line is 60 mV, i.e., $\alpha_2 + \beta_2 = 1$ (region 1).

At high positive potentials another case can also arise corresponding to the 60 mV slope: ionization of H_{ad} is nonactivated, i.e., $\beta_2 = 0$, and electrochemical desorption becomes barrierless ($\alpha_2 = 1$) (region 5).

Regions 2 and 4 correspond to the intermediate situations when one of the reactions is nonactivated and the second proceeds according to the ordinary mechanism with the transfer coefficient equal to ca. $\frac{1}{2}$; for example, in region 2 ionization is activated and electrochemical desorption is not ($\alpha_2 = 0$, $\beta_2 = \frac{1}{2}$). Region 4 corresponds to the opposite case when $\alpha_2 = \frac{1}{2}$ and $\beta_2 = 0$.

In region 3 both reactions are nonactivated and $\log[(2j_0 - j)/j]$ is independent of potential ($\alpha_2 + \beta_2 = 0$). In principle, region 3 can be absent. When the transition from a nonactivated to an activated mechanism of electrochemical desorption takes place at a more negative potential than that corresponding to the transition of activated into nonactivated ionization, then within a certain potential range both reactions proceed with finite activation energies, i.e., $\alpha_2 = \beta_2 = \frac{1}{2}$ and consequently $\alpha_2 + \beta_2 = 1$. This potential range, however, can be too narrow for experimental observation, so that region 2 seems to change over directly into region 4 and the value $\alpha_2 + \beta_2 = \frac{1}{2}$ is preserved over a wide potential range, although the reaction mechanism actually undergoes a change.

Thus, it is obvious that an unambiguous determination of which of the two reactions is nonactivated requires that the experimental $\log[(2j_0 - j)/j]$ vs. φ curve include the region of transition between the sections of the curve presented in Fig. 7.7. However, if the slope of the curve is constant within the potential range studied, as is the case at the mercury electrode (Fig. 7.3), it is equally valid to assume cases corresponding to regions 2 and 4 in Fig. 7.7, i.e., an uninterpreted transition from one mechanism to another.

In order to clarify the picture, it is useful to compare the photoemission data with the experimental results obtained for the case of barrierless discharge of hydrogen ions. The transition from ordinary into barrierless discharge behavior on mercury occurs in the absence of specific adsorption at -0.5 V. In the more negative potential region, discharge of hydrogen ions, and consequently ionization of adsorbed hydrogen, has a transfer coefficient equal to $\frac{1}{2}$. It follows that at more negative potentials, where the photoemission measurements result in $\alpha_2 + \beta_2 = \frac{1}{2}$, electrochemical desorption is nonactivated and ionization proceeds according to the ordinary mechanism.

Assuming, however, that in the region of "hydrogen drop" of the photocurrent the nonactivated process is not the electrochemical desorption but ionization of H_{ad}, it must be expected that the latter must change into the activated process at negative potentials, with $\beta_2 = \frac{1}{2}$. In this case the $j^{0.4}$ vs. φ curve should exhibit an additional increase of photocurrent which is not experimentally observed. Of course, this transition may occur outside the experimentally accessible potential range. Then the linear $j^{0.4}$ vs. φ dependence would mean that both the electrochemical desorption and the ionization steps are nonactivated and the stoichiometric coefficient ν is less than 2. However, Barker and Concialini (1973a) showed that the stoichiometric coefficient of the overall photoemission process is the same when hydrogen ions and N_2O are used as acceptors. In the latter case, the stoichiometric factor is undoubtedly 2. Therefore, it can be assumed that in the potential range below the "hydrogen drop" of the photocurrent, the ionization reaction is virtually absent.

At potentials more positive than -0.7 V (at which the photocurrent starts decreasing on mercury) the rate of ionization of H_{ad} becomes commensurate with the rate of electrochemical desorption, the latter being a nonactivated process. At potentials more positive than -0.5 V (corresponding to the transition from an activated into a barrierless discharge mechanism of H_3O^+ ions on mercury) ionization of atomic hydrogen is nonactivated and the curve φ vs. $\log[(2j - j)/j]$ (Fig. 7.3) should exhibit a limiting current. This has not been found experimentally. It is possible that the limiting current is reached at higher positive potentials where, however, the photocurrent measurements become less accurate owing to the proximity of the photo-electric threshold.

It follows from Fig. 7.3 that at positive potentials the ionization current exceeds by an order of magnitude the current of electrochemical desorption, although the latter is nonactivated, as discussed above. It must be recognized that electrochemical desorption involving hydrogen ions rather than water molecules will proceed at a rate proportional to the product of the ionization rate and the mole fraction of hydrogen ions ($\frac{1}{55}$ in 1 M HCl). Therefore, the relatively low rates of electrochemical desorption indicate that the mechanism involves hydrogen ions.

Finally, the case of the nonactivated to activated transition for the electrochemical desorption process occurring at the same potential at which the activated/nonactivated transition of the ionization reaction takes place should not be excluded. However, the potential at which this change of mechanism might possibly occur could not be experimentally determined.

The above discussion shows that the mechanism of the removal of atomic hydrogen is not yet fully elucidated for mercury. It has, however, been firmly established that one of the coupled reactions of hydrogen removal is nonactivated.

7.5. The Mechanism of Atomic Hydrogen Reactions at Bismuth and Other Solid Electrodes

Although it was not possible to establish which of the coupled reactions (ionization or electrochemical desorption) is nonactivated on mercury, studies of the bismuth electrode have made this distinction possible. The experimental data presented in Fig. 7.4 show a "limiting effect" which appears at potentials above -0.6 V, where the ratio $\overleftarrow{j}_H/\overrightarrow{j}_H$ is virtually independent of potential. This means that both reactions are nonactivated in this potential range (region 3 in Fig. 7.7). Since electrochemical desorption should be nonactivated at more negative potentials, the inflection observed at -0.6 V must be connected with the transition of nonactivated to activated ionization of atomic hydrogen (region 2 in Fig. 7.7). The slope of the line below -0.6 V depends thus on the transfer coefficient, $\beta_2 = 0.4$ (much less than on mercury). The transfer coefficient of the reverse reaction of hydrogen ion discharge on bismuth is close to 0.6 (Palm and Tenno, 1973); i.e., the sum of the transfer coefficients is indeed unity. This agreement with thermodynamic requirements definitely confirms that the transfer coefficient $\beta_2 = 0.4$ refers to the H ionization reaction.

Thus, over the entire potential range studied, electrochemical desorption on bismuth is nonactivated while the ionization reaction is activated when $\varphi < -0.6$ V.

It should be noted that in the region of the "limiting" $\overleftarrow{j}_H / \overrightarrow{j}_H$ value, the ionization current does not exceed the electrochemical desorption current by more than a factor of 2. It follows that electrochemical desorption on bismuth involves water molecules. If the removal of H were to involve hydrogen ions, the ratio of electrochemical desorption to ionization current would have to be (as in the case of mercury) proportional to the mole fraction of hydrogen ions in solution. This conclusion is confirmed by the slight dependence of φ^* on $c_{H_3O^+}$ at bismuth (Fig. 7.6).

The establishment of nonactivated ionization of adsorbed hydrogen on bismuth at potentials above -0.6 V indicates that the barrierless mechanism of H_3O^+ ion discharge applies in this potential range (assuming, as was implicitly done before, that the energy state of H_{ad} produced in the course of the photoemission process is the same as that arising in the cathodic discharge of hydrogen ions). It would be of great interest to compare the photoemission data with those obtained electrochemically on bismuth electrodes. Unfortunately, the literature contains no data on hydrogen overpotential at bismuth electrodes in the potential range discussed.

Such a comparison was made for antimony at which nonactivated ionization and electrochemical desorption of atomic hydrogen are also observed (Rotenberg *et al.*, 1975b). The transition from activated to nonactivated ionization occurs in this case at the same potential (-0.7 V) at which the overpotential curves change slope, the transfer coefficient changing from $\frac{1}{2}$ to ~ 1 (Punning and Past, 1969), i.e., where the discharge of H_3O^+ ions can be assumed to become barrierless.

No photoemission currents were found on platinum in acid solutions (Anson *et al.*, 1976). This is apparently due to the fast oxidation of atomic hydrogen approaching the electrode surface. In fact, the deposition on the hydrogen-covered platinum surface of an excess hydrogen atom results primarily (owing to the equilibrium between adsorbed atomic hydrogen and H_3O^+ ions in solution) in its oxidation. Recombination of adsorbed hydrogen proceeds on platinum with a much slower rate than that of discharge-ionization.

7.6. The Effect of the Double Layer on the Kinetics of Atomic Hydrogen Reactions

Experimental data obtained on mercury and bismuth indicate a considerable effect of specifically adsorbed ions on the rate of removal of atomic hydrogen. This effect is quantitatively accounted for by the introduction of the ψ' potential in the kinetic equations (Delahay, 1965; Frumkin *et al.*, 1952).

The kinetic equations of ionization and electrochemical desorption can then be rewritten as follows:

$$\overleftarrow{j}_H = \overleftarrow{k}_{20} \exp[\beta_2(\varphi - \psi')F/RT]\theta$$
$$\overrightarrow{j}_H = \overrightarrow{k}'_{20} \exp[-\alpha'_2(\varphi - \psi')F/RT]\theta \qquad (7.9)$$
$$+ X\overrightarrow{k}''_{20} \exp(-F\psi'/RT) \exp[-\alpha''_2(\varphi - \psi')F/RT]\theta$$

where α'_2 and α''_2 are the transfer coefficients of the electrochemical desorption involving water molecules and hydrogen ions, respectively. From Eq. (7.9) we have

$$\frac{\overleftarrow{j}_H}{\overrightarrow{j}_H} = \frac{\overleftarrow{k}_{20} \exp\left[\beta_2(\varphi - \psi')\dfrac{F}{RT}\right]}{\overrightarrow{k}'_{20} \exp\left[-\alpha'_2(\varphi - \psi')\dfrac{F}{RT}\right] + X\overrightarrow{k}''_{20} \exp\left(\dfrac{-F\psi'}{RT}\right)\exp\left[-\alpha''_2(\varphi - \psi')\dfrac{F}{RT}\right]}$$
$$(7.10)$$

If it is assumed that the electrochemical desorption is nonactivated and proceeds primarily with water molecules (the case observed on bismuth), then

$$\frac{\overleftarrow{j}_H}{\overrightarrow{j}_H} = \frac{\overleftarrow{k}_{20}}{\overrightarrow{k}'_{20}} \exp[\beta_2(\varphi - \psi')F/RT] \qquad (7.10')$$

A positive shift of the ψ' potential results in a decrease, and a negative shift in an increase, of the $\overleftarrow{j}_H/\overrightarrow{j}_H$ ratio, as was experimentally confirmed on bismuth. In fact, adsorption of Br^- and I^- ions results in an increase of the relative rate of atomic hydrogen oxidation, i.e., $\overleftarrow{j}_H/\overrightarrow{j}_H$ increases (cf. Figs. 7.2 and 7.4).

The effects of halide ions on mercury are qualitatively the same as on bismuth electrodes. The quantitative interpretation, however, is much more difficult for mercury, owing to the participation of hydrogen ions in the overall process on mercury, their local concentration being dependent on the ψ' potential. However, specific adsorption of TBA ions decreases the hydrogen ion concentration in the vicinity of the electrode, and electrochemical desorption then involves only water molecules. Hence, as follows from Eq. (7.10),

$$\frac{\overleftarrow{j}_H}{\overrightarrow{j}_H} = \frac{\overleftarrow{k}_{20}}{\overrightarrow{k}'_{20}} \exp\{[(\alpha'_2 + \beta_2)\varphi - (\alpha'_2 + \beta_2)\psi']F/RT\} \qquad (7.10'')$$

A positive shift of ψ' potential results in a decrease of the ratio $\overleftarrow{j}_H/\overrightarrow{j}_H$ regardless of the value of α'_2. In fact, the experimentally observed "hydrogen drop" shifts toward positive potentials to the extent that it practically disappears.

It should be stressed, once more, that the adsorption of TBA cations causes not only a shift of the ψ' potential but also some blocking of the electrode, preventing the penetration of hydrogen atoms to the surface. However, the latter effect should in the case discussed affect both the cathodic and anodic processes without changing the ratio $\overleftarrow{j}_H/\overrightarrow{j}_H$.

The results described in the last section show that the electrochemical removal of atomic hydrogen is a relatively fast process. Even the data obtained on mercury indicate indirectly the possibility of nonactivated electrochemical desorption. Studies carried out at bismuth electrodes supplied the first experimental evidence for the existence of nonactivated processes in electrochemical kinetics.

It should be mentioned that this concept dates from 1936 when Kabanov carried out the first measurements of hydrogen overpotential at high (up to 100 A-cm^{-2}) current densities in an attempt to discover anomalies in the behavior of the activation energy of the process. However, no definite statement concerning the decrease of the activation energy to zero can be found in that work. Jofa and Mikulin (1944) interpreted (with some reservations) the single point of intersection of extrapolated Tafel lines of hydrogen evolution obtained at various temperatures in terms of zero activation energy. Frumkin, Bagotskii, and Jofa (1951) suggested the existence of the nonactivated electrochemical desorption on mercury. Indirect evidence for nonactivated ionization of hydrogen can be found in Krishtalik's (1965) experiments concerning barrierless discharge. This evidence was based on thermodynamic considerations regarding the sum of transfer coefficients of the forward and back reactions being equal to unity.

Direct measurements of the currents of nonactivated processes are very difficult in practice. Their high rates are always associated with the appearance of concentration polarization. Only recently Erenburg *et al.* (1975) found a system in which the supply of the reactant to the surface was not rate controlling so that the measured limiting current could be treated as that corresponding to the nonactivated process.

The diffusional limitations were successfully overcome in the photoemission studies of atomic hydrogen removal owing to the relative (rather than absolute) nature of the measurements. The latter involved evaluation of the ratio of the rates of two coupled reactions having the same reactant (atomic hydrogen). Diffusional limitations were thus eliminated.

Photoemission As a Method of Investigating Homogeneous Reactions Involving Free Radicals

8.1. Chemical and Electrochemical Reactions of the Radical Anion NO_3^{2-}

The main reactions proceeding in the vicinity of an electrode and on its surface during photoemission of electrons in the nitrate ion solutions can be described as follows:

$$e_{aq}^- + NO_3^- \overset{k_A}{\to} NO_3^{2-} \tag{8.A}$$

$$NO_3^{2-} + H_2O \overset{k_v}{\to} NO_2^- + OH^\cdot + OH^- \tag{8.B}$$

$$OH^\cdot + e^-(M) \to OH^- \tag{8.C}$$

$$NO_3^{2-} \overset{k_{eA}}{\to} NO_3^- + e^-(M) \tag{8.D}$$

(Depending on the actual conditions of the experiment, the above scheme can be modified and supplemented.†) The reaction has been studied in detail on mercury (Eletskii *et al.*, 1970, 1971; Korshunov *et al.*, 1970a; Barker, 1971) and on solid metals (Rotenberg *et al.*, 1974, 1975b). Equations (8.A)–(8.D) correspond to the general scheme discussed in Section 2.2, in which the product [eA] (in the present case NO_3^{2-}) becomes oxidized and the product [R] (in the present case the $OH^\cdot$ radical) becomes reduced at the electrode. The resulting photocurrent is then given [cf., e.g., (2.24)] by

$$j = \nu\left\{I - I_e - \frac{e}{1 + (Q_v \mathscr{D}'/k_{eA})} \frac{Q^2}{Q^2 - Q_v^2} \int_0^\infty \Phi(x)(e^{-Q_v x} - e^{-Qx})\,dx\right\} \tag{8.1}$$

where I_e is the reverse current for the return of hydrated electrons to the

† Another possible mechanism discussed by Barker *et al.* (1970) is
$$NO_3^{2-} + H_2O \to NO_2 + 2OH^-$$
$$NO_2 + e^-(M) \to NO_2^-$$
$$NO_3^{2-} \to NO_3^- + e^-(M)$$
This change in mechanism will not affect the validity of the following discussion.

electrode, $Q \equiv (k_A c_A / \mathscr{D}_e)^{1/2}$, $Q_v \equiv (k_v \mathscr{D}')^{1/2}$, and $\mathscr{D}'$ is the diffusion co-efficient of NO_3^{2-} radicals. The indices on rate constants follow Eqs. (8.A)–(8.D). The stoichiometric coefficient ν is equal to the number of electrons transferred across the interface for each decomposed NO_3^{2-} ion. For reactions (8.A)–(8.D), $\nu = 2$.

Equation (8.1) assumes that "dry" electrons are not captured by the acceptor. It should be noted that according to Hart and Anbar (1970) the nitrate ion can react with dry electrons. In this case Eq. (8.1) would need some modification. A more detailed discussion of this problem can be found in a paper by Pleskov and Rotenberg (1974).

The value of ν can be experimentally determined in each case from a comparison of the photocurrent observed in the nitrate-containing solution with that observed for another acceptor having a known stoichiometric coefficient. Thus, at negative potentials at which the oxidation of NO_3^{2-} can be neglected the photocurrent is about 1.5 times higher in neutral nitrate solutions than in acids (cf. Section 5.2). Since the firmly established value of the stoichiometric coefficient for hydrogen ions is 2, its value for nitrate ions must be 3. The latter value, higher than that which follows from the simplified reaction scheme (8.A)–(8.D), indicates that the latter must be supplemented by a step involving reduction of NO_2^- ions:

$$NO_2^- + e(M) + H_2O \rightarrow NO + 2OH^- \qquad (8.E)$$

Thus, one photoelectron induces an additional transfer across the interface of two electrons by means of electrochemical reactions.

It follows from Eq. (8.1) that two potential regions exist in which the potential dependence of the photocurrent reflects only the emission current behavior. The negative potential region corresponds to $Q_v \mathscr{D}' / k_{eA} \gg 1$ and consequently $j = \nu(I - I_e)$, and the positive one to $Q_v \mathscr{D}' / k_{eA} \ll 1$. In the latter case for $Q_v x_0 \ll 1$, $Q x_0 \gg 1$, we obtain

$$j = \nu x_0 Q_v I \qquad (8.2)$$

In the intermediate region the expression for the photocurrent explicitly contains the rate constant of the electrode reaction, k_{eA}, which itself is potential dependent: $k_{eA} = k_{eA}^0 \exp(\beta \varphi F / RT)$. This feature of the potential dependence of the photocurrent in the intermediate region forms the basis of the quantitative method of investigation of the electrochemical oxidation of NO_3^{2-} ions.

Experimental data obtained on mercury are shown in Fig. 8.1 in the form of the dependence of $j^{0.4}$ vs. φ. At high NO_3^{2-} concentrations ($> 0.01\ M$), S-shaped curves are obtained: Two linear sections are separated by a region ($-1.2\ V < \varphi < -0.7\ V$) of sharp variation of the photocurrent with potential.

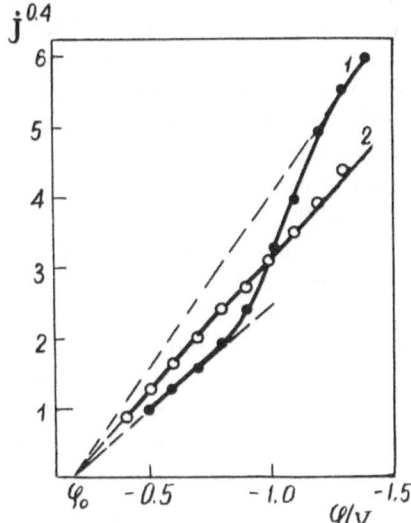

Fig. 8.1. Reactions of NO_3^{2-} (Eletskii *et al.*, 1971). Dependence of $j^{0.4}$ on φ at a mercury electrode in 1 *M* KNO_3, in the absence (1) and the presence (2) of TBA.

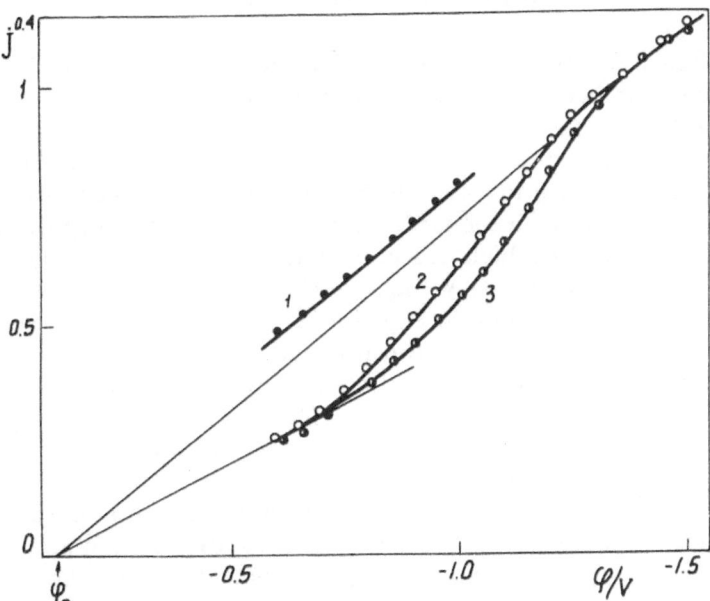

Fig. 8.2. Reactions of NO_3^{2-} (Rotenberg *et al.*, 1974). Dependence of $j^{0.4}$ on φ in 1 *M* KNO_3. Electrodes: (1) indium; (2) bismuth; (3) lead.

Similar relations were observed at bismuth and lead electrodes (Fig. 8.2), whereas on indium the decrease of photocurrent, interpretable in terms of NO_3^{2-} oxidation, has not been observed. This may be due to the catalytic decomposition of NO_3^{2-} (caused by indium oxides present at the electrode surface) which proceeds much faster than electro-oxidation. The effect of surface oxides on the rates of decomposition and oxidation of the NO_3^{2-} radical ion has been also described for platinum electrodes (Rotenberg et al., 1975b).

Comparison of the photocurrent observed at high positive potentials [at high acceptor concentrations when Eq. (8.2) is obeyed] with the emission photocurrent determined by interpolation between the point corresponding to the threshold potential, φ_0, and the section of the $j^{0.4}$ vs. φ curve obtained at negative potentials allows, in principle, the rate constant k_v of the homogeneous NO_3^{2-} decomposition to be determined. It must be stressed that this method of determining k_v is valid only if the "limiting" current for NO_3^{2-} electro-oxidation in the region of more positive potentials is determined by a diffusional rather than a kinetic step. The rate constants k_v found in this way for various metals are of the same order of magnitude, 10^6–10^7 sec^{-1} (Eletskii et al., 1970, 1971).

The value of k_v can also be determined from transient photocurrents induced by pulse illumination of electrode. In the latter case k_v is evaluated from the kinetics of the rise of photopotential with time due to the formation of $OH^{\cdot}$ radicals from reduction of decomposition products of NO_3^{2-}. The k_v values thus obtained are 10^5 sec^{-1} (Barker, 1971) and $4 \cdot 10^5$ (Zolotovitskii et al., 1975).

Apart from the rate constants for NO_3^{2-} decomposition, photoemission measurements can help to establish the rate constant of electro-oxidation of this radical ion and its dependence on potential. For this purpose, Eq. (8.1) must be simplified. At high acceptor concentrations ($Qx_0 \gg 1$), if simultaneously $Q_v x_0 \ll 1$, we have $k_{eA}/Q_v \mathscr{D}' = (\nu I - j)/(j - j_1)$ where $j_1 = \nu x_0 Q_v I$. Taking into account the exponential dependence of k_{eA} on φ, a linear relation should be obtained between φ and $\log[(\nu I - j)/(j - j_1)]$. Experimental data plotted in the latter coordinate system allow the transfer coefficient β of the electro-oxidation of NO_3^{2-} ions to be determined. For mercury, $\beta = 0.25$ (Eletskii et al., 1971) while for lead and bismuth electrodes, $\beta = 0.35$ (Rotenberg et al., 1974).

The rate of NO_3^{2-} oxidation is strongly affected by specific adsorption, for example, of iodide and TBA ions. The effect of TBA ion adsorption on mercury is shown in Fig. 8.1. At negative potentials the photocurrent is lower, and at positive potentials higher, in the presence than in the absence of TBA. This effect of TBA cations [observed also on bismuth and lead (Rotenberg et al., 1974)] is connected, on the one hand, with the inhibition of photoemission itself, and on the other, with inhibition of electro-oxidation of NO_3^{2-}. Large TBA cations evidently block access of NO_3^{2-} ions to the surface, resulting in the decomposition of the latter in solution. Therefore the "oxidative" photocurrent drop disappears in the presence of TBA.

8.2. Chemical and Electrochemical Reactions Involving CO_2^- and CH_3^* Radicals

The redox reaction in solution of the CO_2^- radical anion is similar in nature to the process involving NO_3^{2-}. The CO_2^- radical is formed upon the capture of a hydrated electron by a CO_2 molecule. Photoemission of electrons into solutions saturated with carbon dioxide has been studied in detail by Schiffrin (1972, 1974). The stationary photocurrent is observed only at potentials below -1.4 V, where the radical anion becomes reduced in the presence of proton donors to the formate ion:

$$CO_2^- + H_3O^+ + e^-(M) \rightarrow HCOO^- + H_2O \qquad (8.F)$$

At more positive potentials virtually all CO_2^- radicals are oxidized at the electrode and the photocurrent disappears.

The potential dependence of the stoichiometric coefficient $\nu(CO_2)$ of the overall photoprocess is shown in Fig. 8.3. The "half-wave potential" φ^* of the electrode reaction of CO_2^- can be determined from this plot. At φ^* the rates of oxidation and reduction of CO_2^- are equal (cf. Section 7.2) and $\nu(CO_2^-) = 1$. The value of φ^* is independent of pH above pH $= 6.5$. This led Schiffrin to conclude that when pH > 6.5 the proton donors are water molecules. In acid solutions φ^* depends considerably on pH, indicating the participation of H_3O^+ ions in the reduction of CO_2^-.

The CH_3^* radical is formed as a result of capture of a hydrated electron by CH_3Cl. As opposed to CO_2^-, the CH_3^* radical is not oxidized at mercury but undergoes dimerization to ethane. Therefore finite photocurrent values are observed over the entire potential range studied. It should be mentioned

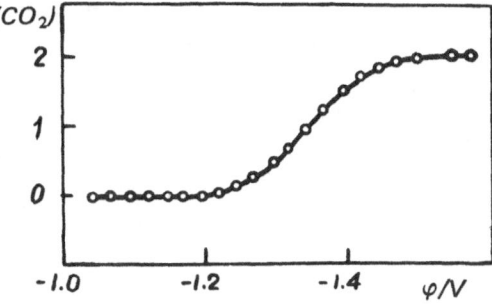

Fig. 8.3. A study of CO_2^- reactions (Schiffrin, 1974). The stoichiometric coefficient $\nu(CO_2)$ as a function of potential. Mercury electrode, 0.05 M KHCO$_3$ + 0.95 M KCl saturated with N$_2$O.

that the reduction of aliphatic monohalide derivatives was never observed polarographically in aqueous solutions.

8.3. Chemical Reactions of H· and OH· Radicals with Alcohols and Other Organic Compounds

Photocurrents observed in solutions of acceptors such as N_2O and H_3O^+ are considerably decreased in the presence of alcohols (methanol, ethanol, etc.). This is due to homogeneous chemical reactions between alcohols and electron-capture products [eA]. The radicals H· and OH· formed in acid solutions of N_2O remove the α-hydrogen atom from alcohol molecules with relative ease. The resulting RĊHOH radical is oxidized over a wide potential range to aldehyde, whereas the original radical (OH· or H·) is reduced at these potentials. The consecutive steps initiated by photo-emission in such systems can be represented by the general scheme in Table 8.1. According to this scheme, two products participate in electrochemical reactions: [eA](=OH· or H·) and [R](=RĊHOH). In the potential range studied, the first product is reduced, and the second is reduced or oxidized at the electrode depending on the electrode potential. This case is described by Eq. (2.27). At potentials corresponding to oxidation of the RĊHOH radical, the photocurrent decreases in the presence of alcohols (Barker, 1968).

The effect of methanol on the photocurrent arising in a solution saturated with N_2O (10%) + N_2 is shown in Fig. 8.4. Above $-1.4\,V$ a small addition of methanol (0.05 M) already results in a considerable decrease of the photocurrent.

In this potential region, all ĊH$_2$OH radicals are electro-oxidized. The sharp increase of the photocurrent at about $-1.5\,V$ is connected with the reduction of ĊH$_2$OH to methanol which commences at this potential.

Table 8.1. Steps Initiated by Photoemission in Some Solutions of Acceptors

N_2O Solutions	H_3O^+ Solutions
$N_2O + e_{aq}^- + H_2O \rightarrow N_2 + OH\cdot + OH^-$	$H_3O^+ + e_{aq}^- \rightarrow H\cdot + H_2O$
$OH\cdot + e(M) \rightarrow OH^-$	$H\cdot + e^-(M) + H_2O \rightarrow H_2 + OH^-$
$OH\cdot + RCH_2OH \overset{k_Y^0}{\rightarrow} R\dot{C}HOH + H_2O$	$H\cdot + RCH_2OH \overset{k_Y^0}{\rightarrow} R\dot{C}HOH + H_2$

$$H_2O + R\dot{C}HOH \overset{k_R}{\rightarrow} RCHO + e^-(M) + H_3O^+$$
$$R\dot{C}HOH + e^-(M) + H_2O \rightarrow RCH_2OH + OH^-$$

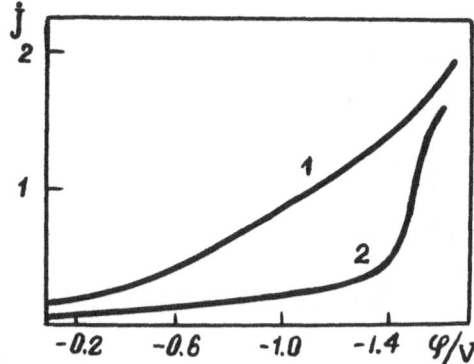

Fig. 8.4. Reaction of methanol with the hydroxyl radical (Barker, 1968). Photocurrent observed in 0.2 M KCl solution saturated with N_2O (10%) + N_2 in the absence (1) and the presence of 0.05 M methanol (2).

The kinetics of the photoprocess studied under conditions of pulse illumination led Barker (1968) to conclude that the rate of the anodic process is controlled by oxidation of alcohol radicals adsorbed at the electrode surface. The experimental rate constants for $CH_3\dot{C}HOH$ oxidation depend little on potential, possibly owing to slow adsorption preceding the electrochemical step.

The quantitative relation between the photocurrent and the concentration of alcohol can be found from Eq. (2.28) which is valid in the case of complete oxidation of $R\dot{C}HOH$ radicals:

$$j = j_{k_v=0} \frac{Q}{Q + Q_v} \tag{8.3}$$

where $j_{k_v=0}$ is the photocurrent in the absence of alcohol, $Q_v = (k_v/\mathscr{D}')^{1/2}$, $k_v = k_v{}^0 c_{\text{alc}}$, c_{alc} is the concentration of the alcohol in solution, and $\mathscr{D}'$ is the diffusion coefficient of $H\cdot$ or $OH\cdot$. The above equation, implying a linear dependence of j^{-1} on $c_{\text{alc}}^{1/2}$, was used by Barker (1968) to determine the rate constants for reactions of alcohols (ethanol and methanol) with hydrogen atoms and $OH\cdot$ radicals. The experimental data for ethanol in acid solutions are shown in Fig. 8.5 as a j^{-1} vs. $(c_{C_2H_5OH})^{1/2}$ plot. Such a plot allows the rate constant $k_v{}^0$ of the reaction of atomic hydrogen with alcohol to be determined. Both rate constants, for methanol and ethanol, have the same value: $2.4 \cdot 10^7$ mole^{-1}-liter-sec^{-1}. The rate constant of the reaction of the $OH\cdot$ radical with the same alcohols is 10^9 mole^{-1}-liter-sec^{-1}. These values are in very good agreement with the data obtained from pulse radiolysis of alcohol solutions.

The same method was used by Eletskii and Pleskov (1974) in their studies of $OH\cdot$ reactions with higher alcohols, and by Barker and Bottura (1973a) who investigated photoemission into formate ion solutions. (The

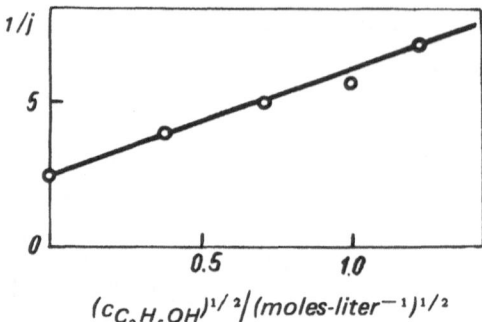

Fig.8.5. Photocurrent in the presence of alcohols plotted in terms of Eq. (8.3) (Barker, 1968). Solution: 0.2 M KCl + 10^{-3} M HCl with addition of ethanol.

latter ions react with OH to form CO_2^- radical anions which, as has already been mentioned, become oxidized at the electrode.) Alcohol changes not only the voltammetric characteristics of the photocurrent, but also the character of the dependence of photocurrent on concentration of acceptors such as N_2O, H_3O^+, etc. It can be shown [Eq. (2.27)] that if the $R\dot{C}HOH$ radicals do not become oxidized at the electrode, the relation $j = f(c_A^{1/2})$ holds both at a low acceptor concentration and in the absence of alcohol. Under conditions of oxidation of $R\dot{C}HOH$ for $Q_v \gg Q$, the photocurrent is linear with c_A. Both types of dependence of photocurrent on hydrogen ion concentration in the presence of methanol are shown in Fig. 8.6. Curve 1 is obtained at -1.0 V where $CH_3\dot{C}HOH$ undergoes oxidation at the electrode and $\overleftarrow{j} = f(c_{H_3O^+})$, while curve 2 is obtained at -1.4 V where the $CH_3\dot{C}HOH$ radical undergoes reduction and $\overrightarrow{j} = f(c_{H_3O^+}^{1/2})$. The agreement between the observed behavior and theory provides additional experimental evidence

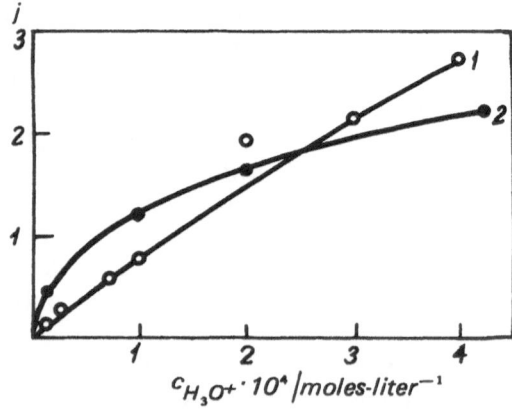

Fig. 8.6. Photocurrent as a function of acceptor (H_3O^+) concentration in solutions containing a high concentration of ethanol. Mercury electrode, 0.2 M KCl + 1 M C_2H_5OH. Potential: -1.0 V (1); -1.4 V (2).

for the validity of the suggested mechanism of the effect of alcohols on photodiffusion currents.

The OH· radicals can split not only α-hydrogen of alcohol molecules but also other hydrogen atoms. A mixture of "secondary" radicals is then formed, and the determination of the "photoemission yield" of each of them is, in general, rather difficult. Barker and Bolzan (1974a, b) suggested a method for evaluation of the yield of α-radicals in the overall process, (i.e., radicals resulting from splitting off α-hydrogen atoms) based on the difference of electro-oxidation rates of various radicals (α-radicals are oxidized at higher rates).

If it is assumed that at a given potential all radicals except "α" are oxidized at a very low rate, then under stationary conditions, when Eq. (8.3) is obeyed, all radicals eventually undergo oxidation and the separate contributions cannot be determined. However, under transient conditions, e.g., in pulse illumination, slow reactions cannot occur significantly within a given period of time τ, while the fast electro-oxidation of α-radicals can be completed. In this case, the photocurrent at the time τ exceeds the stationary value, and its magnitude allows the determination of the yield of α-radicals.

Barker and Bolzan (1974a) displayed their experimental data as plots of Z vs. $c_{\text{alc}}^{-1/2}$, where $Z = j_{k_v = 0}/(j_{k_v = 0} - j)$. Under stationary conditions, when all radicals undergo oxidation, $Z = Q/Q_v$ [cf. Eq. (8.3)] and the Z vs. $c_{\text{alc}}^{-1/2}$ line passes through the origin. Under nonstationary conditions, when at a time τ only a fraction ξ of radicals is oxidized (e.g., only α-radicals), the following expression obtains:

$$Z = (1 - \xi)/(1 + \xi) + 2Q/(1 + \nu_{\text{R}})Q_v \qquad (8.4)$$

where ξ is a quantity obtainable by solving a nonstationary diffusion equation for each type of radical (Barker and Bolzan, 1974a). For $\xi < 1$, a linear Z vs. $c_{\text{alc}}^{-1/2}$ plot is obtained with an intercept on the Z coordinate of

$$Z_\infty = (1 - \xi)/(1 + \xi) \qquad (8.5)$$

If α-radicals are oxidized at a much faster rate than others, then at a suitable potential and after a sufficiently long time τ, the quantity $\xi = Y_\alpha$, where Y_α is the fraction of the α-radicals in the overall reaction.

Figure 8.7 exemplifies the approach described above. It shows the Z vs. $c_{\text{alc}}^{-1/2}$ plots obtained for 1-propanol in a solution containing N_2O. The reaction of 1-propanol with OH· radicals results in three other types of radicals. In the -1.5 to -1.3 V potential range, the experimental points lie on a single straight line (1) which upon extrapolation to $c_{\text{alc}}^{-1/2} = 0$ yields $Z = 0.4$. Assuming that only α-radicals are oxidized in this potential range,

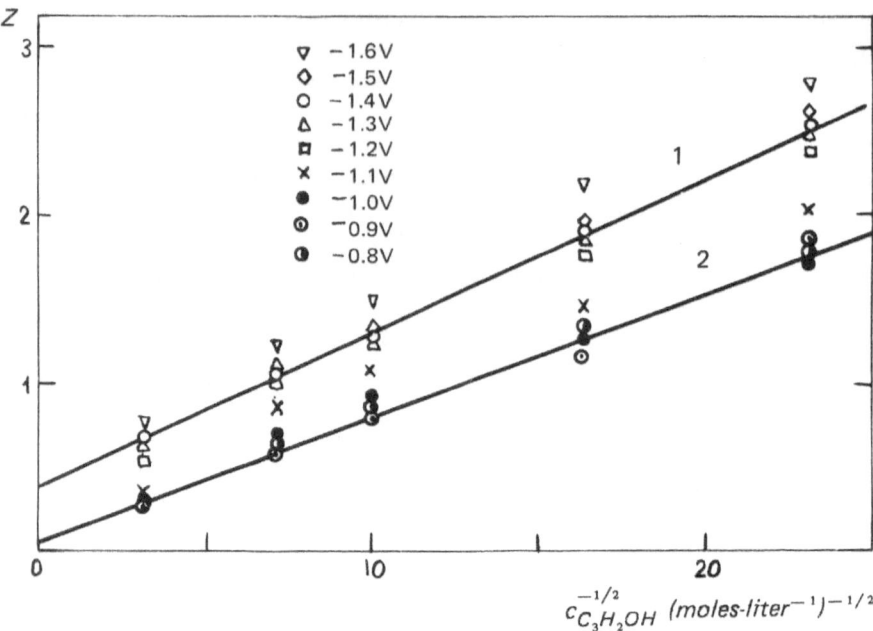

Fig. 8.7. The Z vs. $c_{C_3H_7OH}^{-1/2}$ plots obtained in the presence of 1-propanol (Barker and Bolzan, 1974b). Solution: 0.2 M KCl saturated with N_2O (10%) + Ar. Potential values as shown in the figure.

their relative yield determined using Eq. (8.5) is 0.43. At potentials above -1.1 V, the experimental points obtained at various potentials are grouped along the straight line (2) passing through the origin. Consequently, in this potential range, all radicals undergo oxidation, i.e., $\xi = 1$, and Eq. (8.3) applies, which allows the rate constant of the reaction of OH· radicals with the alcohol to be determined.

In a more general case, when the rates of oxidation of various radicals are commensurate and the Z vs. $c_{alc}^{-1/2}$ plots do not extrapolate to the origin, experimental data can be plotted as W vs. $c_{alc}^{-1/2}$, where $W = (Z - Z_\infty)/(1 + Z_\infty)$. This treatment applied to photocurrents observed in solutions of various alcohols led to the determination of the rate constants of their reactions with H· and OH· radicals (from the slopes of the straight lines obtained), as shown in Table 8.2.

An even more complex system, "acceptor + two organic substances," was discussed by Bottura et al. (1975). Photoelectron emission into such a solution involves competition between the organic substances (e.g., methanol

Table 8.2. **Kinetic Characteristics of the Reactions of** H· **and** OH· **Radicals with Alcohols**

Radical	Alcohol	$k_v{}^0 \cdot 10^{-9}$ mole^{-1}-liter-sec^{-1}		Y_α	
		Photo-emission[a]	Radio-chemical[b]	Photo-emission[a]	Radio-chemical[c]
OH·	Methanol	0.63	0.45–0.65		
	Ethanol	1.2	1.0–1.85	0.9	0.97
	1-Propanol	0.9	1.7	0.43	0.61
	2-Propanol	1.0	1.1–1.7	0.85	0.89
	1-Butanol	1.5	2.2	0.4	0.34
	tert-Butanol	0.3	0.25–0.42		
	1-Pentanol	1.8		0.4	
H·	Methanol	0.024			
	Ethanol	0.024			

[a] Barker and Bolzan, 1974b; [b] Adams and Wilson, 1969; [c] Anbar and Neta (1967).

and phenol) in their reactions with the primary radicals (e.g., in N_2O solutions, OH·) formed by the capture of hydrated electrons by the acceptors. The method developed by Bottura *et al.* (1975) resembles somewhat that described in Section 5.6 using competing acceptors. Using their method and knowing the rate constant of the reaction of OH· with methanol, these authors determined the rate constant for phenol.

Section 8.3 has served basically to illustrate certain results obtained in a large series of papers by Barker and his co-workers. The photoemission method, it is seen, can yield extensive information concerning the kinetic characteristics of homogeneous reactions, thanks to the existence of the rather universal theory of photodiffusion currents for such systems.

8.4. Multielectron Electrochemical Reactions Initiated by Photoemission

Examples considered in the previous text involved usually one-electron electrochemical reactions of the capture products [eA]. However, the most efficient use of light energy would seem to require systems in which photoemission initiates multielectron electrochemical reactions. Therefore, investigation of the photoemission processes in aqueous SF_6 solutions (Concialini *et al.*, 1974) are of considerable interest. Although chemically inactive, SF_6 is an efficient acceptor of hydrated electrons (Asmus and Fendler, 1968). Unfortunately, SF_6 has a very low solubility in water and the photoemission measurements are limited to the low concentration range ($c_A \ll 10^{-3} M$).

Qualitatively, in the -1.0 to -1.6 V range the dependence of its photocurrent on potential at a mercury electrode in SF_6 solutions is the same as in N_2O solutions. At higher potentials (> -0.9 V), the photocurrent sharply decreases with polarization, owing to the incomplete electroreduction of the electron capture products. The potential dependence of the stoichiometric coefficient $\nu(SF_6)$ for photoemission into SF_6 solutions is shown in Fig. 8.8. It was determined by comparison of photocurrents observed in SF_6 ($c_{SF_6} = 2.2 \cdot 10^{-4} M$) and N_2O ($c_{N_2O} = 2.4 \cdot 10^{-3} M$) solutions using a relation derived from Eq. (2.14):

$$\frac{j(SF_6)}{j(N_2O)} = \frac{\nu(SF_6)}{\nu(N_2O)} \left(\frac{k_{SF_6}c_{SF_6}}{k_{N_2O}c_{N_2O}}\right)^{1/2}$$

The stoichiometric coefficient $\nu(N_2O)$ was assumed to be 2 and the rate constants of reactions with e_{aq}^- obtained in radiation-chemical measurements were used. At more negative potentials, the $\nu(SF_6)$ value reaches a limiting value, close to 8. A similar $\nu(SF_6)$ value was obtained using the competing acceptor method with hydrogen ions. At more positive potentials, $\nu(SF_6)$ sharply decreases. The limiting value of $\nu(SF_6)$ at positive potentials could not be determined owing to the low accuracy of measurements at small $\nu(SF_6)$ and c_{SF_6} values.

The mechanisms of reactions of the electron capture products are similar for SF_6 in both radiolysis of its aqueous solutions and in photoemission. At least, the first two steps are identical in both processes:

$$SF_6 + e_{aq}^- \rightarrow SF_6^- \rightarrow SF_5^{\cdot} + F^-$$
$$SF_5^{\cdot} + 2H_2O \rightarrow OH^{\cdot} + H_3O^+ + F^- + SF_4$$

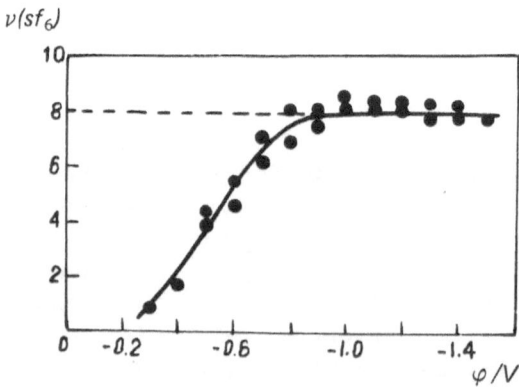

$\nu(sf_6)$

Fig. 8.8. Stoichiometric coefficient $\nu(SF_6)$ as a function of potential (Concialini et al., 1974). Mercury electrode, 0.2 M KCl.

(The formation of OH˙ radicals in the overall photoprocess is confirmed by measurements of photocurrents obtained in the presence of ethanol. The principle of the method is described in Section 8.3.) Upon radiolysis, SF_6 undergoes further chemical reactions, whereas electrochemically it is reduced as follows:

$$SF_4 + 6e^-(M) \rightarrow S^{2-} + 4F^-$$

Since OH˙ radicals are also reduced at the electrode, the overall stoichiometric coefficient $\nu(SF_6) = 8$.

Finally, it should be mentioned that SF_6 solutions are the first for which high ν values have been obtained. The low solubility of SF_6 in water does not, however, allow a high efficiency of light energy conversion to be attained. The problem can probably be solved by a search for other solvents in which the solubility of SF_6 is greater.

Photoelectron Emission from Semiconductors into Solutions and from Solutions into the Vapor Phase

9.1. Qualitative Description of Photoemission from Semiconductors

The two characteristic frequency regions, the near threshold and the extrathreshold (see Chapter 1), of the incident light have already been mentioned in connection with photoelectron emission from metals described in Section 1.1. The photoemission laws in the extrathreshold region are closely connected with the properties of the emitting metal, whereas the near-threshold region supplies rather general information largely independent of the metal structure. Physically, the possibility of obtaining the latter type of information originates from the fact that a large number of electron-filled states exist in the conduction band of the metal.

Photoemission from semiconductors and dielectrics represents a more complex case. On the one hand, the specificity of the behavior of electromagnetic fields penetrating into semiconductors and dielectrics affects the character of the photoemission. On the other hand, basic complications arise owing to the rather important role played by collective effects, even at relatively low excitation energies. Nevertheless, the general approach developed in Chapter 1 can be utilized in the discussion of photoemission from semiconductors and, in particular, from semiconductors in contact with electrolytes.

Similar to the case of metals, the analysis of the optical properties of semiconductors can be made by consideration of separate frequency regions (Willardson and Beer, 1967). The first region is the far infrared part of the spectrum; the light absorption here is rather weak, being mainly due to the interaction of light with lattice vibrations. The second region corresponds to the visible and near-visible regions of the spectrum. Its characteristic feature is a well-defined structure due to the absorption caused primarily by electron transitions into the conduction band from the valence band and possibly also from various energy levels in forbidden bands. The third region is characterized by decreasing absorption and a sharp drop of reflectivity, resembling the behavior of metals in the region of ultraviolet transparency. The third region corresponds to frequencies $\omega > \omega_p$, the value of ω_p being evaluated from the same expression as in the case of metals (p. 11), in which n_e denotes in the present case the electron concentration in the valence band of the semiconductor. Further increase of the quantum energy (up to energies of the order of several tens of electron volts) results in the re-appearance of considerable optical absorption, due mainly to the electron transitions into the conduction band from deep inner filled bands.

Photoemission from semiconductors is energetically possible at frequencies belonging to the second region. This very region is of primary interest in considering photoemission in the near-threshold frequency range.

An important feature of electron photoemission from semiconductors arises from the fact that the chemical potential μ which determines the value of the thermodynamic work function $w^{(\mathrm{th})}$ is usually located in the forbidden band. (The exceptions are heavily doped semiconductors, which are beyond the scope of the present discussion.) Therefore, electrons with initial energy $E_i = \mu$ are absent in semiconductors and the minimum energy $\hbar\omega_0$, which corresponds (at $T = 0$) to the energetically possible single-photon emission, differs from the thermodynamic work function (cf. Fig. 9.1). The photo-emission work function $w^{(\mathrm{ph})} = \hbar\omega_0$ is obviously equal to $\hbar\omega_0 = E_g + \chi$, where E_g is the width of the forbidden band and χ is the difference between the potential energy level of delocalized electrons outside the emitter and the energy level of the bottom of the conduction band in the crystal; it is called the electron affinity of the semiconductor.

The values of $w^{(\mathrm{ph})}$ and χ depend not only on the bulk properties of the crystal, but also on the properties of the medium and the interface. Owing to the polarization interaction of emitted electrons with the medium and to the existence of additional dipole potential drops at the interface (due, e.g., to adsorption), the magnitude of χ can be quite small and, in certain cases, even negative. In particular, the case $\chi < 0$ can be achieved in practice by adsorption of Cs and its compounds at the vacuum interface of p-semi-conductors. The quantum yield of photoemission from emitters with a nega-tive electron affinity is often close to unity. Emitters of this type have been widely studied, owing to their potential practical applications (Laar and Scheer, 1968; Bell and Spicer, 1970; Petrov, 1971; Soboleva, 1973). Studies of photoemission from semiconductors into polar media (electrolytes), which in contact with the emitter further decrease its electron affinity, prove to be rather fruitful in this respect.

Photoexcitation of electrons from ideal semiconductor crystals at $T = 0$ can, according to the single-particle model, proceed by surface or

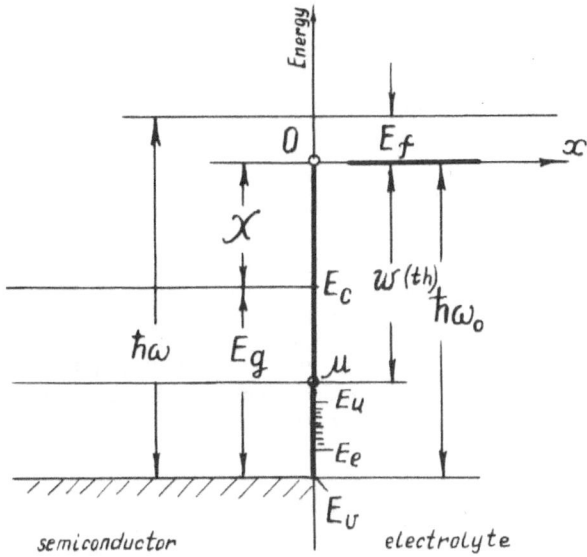

Fig. 9.1. Energy diagram for photoemission from a semiconductor electrode.

bulk generation mechanisms. As opposed to the case of metals, bulk photo-excitation, which usually plays the main role in the case of semiconductors, is necessarily accompanied by interband transitions, the reason being that negligibly low concentrations of electrons exist in the conduction band, so that photoexcitation occurs almost exclusively from filled bands.

Interband transitions are customarily divided into the direct and indirect ones. Direct bulk transitions, which in the band diagram (cf. Fig. 9.2) are "vertical," consist in the interaction of a single electron having an initial energy E_i and a photon of energy $\hbar\omega$. This results in the transition of the electron to the second band. Since the photon momentum $\simeq \hbar\omega/c$ (c is the light velocity) is negligibly small, the quasi-momentum of the electron in the crystal does not change; thus $\mathbf{p}_i = \mathbf{p}_f$, where $\mathbf{p}_f = \{p_{fx}, \mathbf{p}_{\|}\}$ is the momentum of a photoexcited electron in the conduction zone and $\mathbf{p}_i = \{p_{ix}, \mathbf{p}_{\|}\}$ is the initial momentum of the electron in the valence band. Owing to the existence of the surface potential barrier ($\chi \neq 0$), $\mathbf{p}_f$ differs from p, the final momentum of the electron outside the emitter, $\mathbf{p} = \{p, \mathbf{p}_{\|}\}$. Not only momentum, but energy, must also be conserved, i.e.,

$$E_f(\mathbf{p}_f) - E_i(\mathbf{p}_i) = \hbar\omega \qquad (9.1)$$

where energies E_f and E_i are functions of the quasi-momentum in the respective bands.

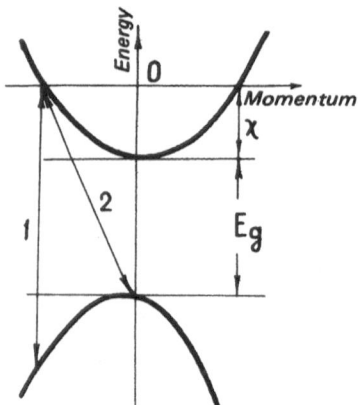

Fig. 9.2. Diagram of interband transitions in a semiconductor. (1) Direct transition; (2) indirect transition.

Indirect transitions involve interactions with one or several phonons, as well as with trace impurities, vacancies, etc. It is obvious that in this case $\mathbf{p}_i$ differs in general from $\mathbf{p}_f$. In the general energy balance, however, the phonon energy can usually be neglected and trace impurities and vacancies do not transfer energy to electrons. Therefore the energy conservation law written in the form (9.1) can apply also in the latter case. The indirect transitions are obviously less probable in high-purity crystals than the direct ones.

It can be seen from Fig. 9.2 that the electron can acquire the energy $E_f > 0$, necessary for photoemission, in a direct transition if the energy of the absorbed photon is larger than $E_g + \chi$. In other words, the generation of electrons in the bulk semiconductor according to the mechanism of direct transitions when other mechanisms are insignificant has a threshold energy $\hbar\omega_0'$ higher than $\hbar\omega_0 = w^{(ph)}$ (in Fig. 9.2 the energy $\hbar\omega_0'$ corresponds to the length of the arrow 1). The difference $\omega_0' - \omega_0$ depends on the dispersion law in bands and on the orientation of the emitter's surface with respect to the crystallographic axes. However, in the absence of limitations due to the law of conservation of quasi-momentum in interband transitions (i.e., in the case of surface excitation), the threshold frequency is solely determined by the requirements of the energy conservation law, and thus $\omega_0 = w^{(ph)}/\hbar$.

Thus, the threshold frequency of photoemission from semiconductors depends on the mechanism of photoelectron generation.

This qualitative description is valid with one reservation: The translational symmetry of a crystal is distorted in the x direction where it occupies a half-space only; therefore the x component of the quasi-momentum is not conserved, strictly speaking, under any conditions (conservation of the tangential components usually follows from the translational symmetry in the directions y and z). Correspondingly, "strictly direct" transitions should not be considered. Nevertheless the relation $p_{ix} = p_{fx}$ is highly accurate if the electrons are excited at a sufficient depth in the crystal. In mathematical terms: $d_x/l \ll 1$, where d_x is the lattice constant in the x direction and the quantity l is determined in the simplest case by the smallest of the following characteristic dimensions: the distance l_ω of the light wave field attenuation in the semiconductor, and the path length of the photoexcited electron with respect to inelastic, l_i, and elastic, l_e interactions: $l = \min\{l_\omega, l_i, l_e\}$. If the condition $d_x/l \ll 1$ is not satisfied, the concept of direct transitions is meaningless and the picture of photoexcitation described above must be modified.

An important difference between photoelectron emission from semiconductors and metals results from the law of change of density of electron states in the band which supplies emitted electrons. For metals, the initial electron states, which contribute to photoemission in the near-threshold region, lie in the vicinity of the Fermi level. The density of states is a slowly changing function of energy. For semiconductors, the density of states becomes zero at energies corresponding to the photoelectric threshold, since the highest filled state of energy E_i coincides with the top of the valence band, E_v. This also leads to the changed character of the functional dependence of the photoemission current I on the light frequency ω.

Photoemission from semiconductors can be considerably affected by the surface states (Davison and Levine, 1970; Many *et al.*, 1965). If the energies of electrons localized at the surface lie in the forbidden band, the photoelectric threshold shifts toward lower quantum energies (in metals the surface states play a negligible role in photoemission owing to the large number of electrons in the conduction band). Also, the presence of charge on surface states (as well as the superposition of the external field) results in the deformation of the energy bands at the semiconductor interface and affects the laws of the process considered.

The propagation of the photoelectron in the bulk crystal toward the emitting surface is the central problem in the theoretical analysis of photoelectron emission under conditions of bulk excitation. It is often assumed (e.g., James and Moll, 1969; Baksht *et al.*, 1971) that this propagation can be described in a classical way by introducing these or other modifications in the treatment of the random walk of a point center in a medium. Another approach (Kane, 1962; Gurevich, 1972) considers the propagation of the probability wave described by suitable quantum-mechanical equations. The complete quantum-statistical description contains, in principle, both approaches as limiting cases. However, a consistent mathematical realization of such limiting transitions turns out to be a very complex problem (cf. Langreth, 1971).

The two limiting cases correspond to two different physical situations. The first approach is applicable if the majority of photoelectrons undergoes a considerable number of interactions. The second approach is valid when the emitted electrons escape from the crystal with virtually unchanged quantum characteristics acquired during photoexcitation (e.g., at light frequencies close to the threshold and, besides, with $l_\omega < l_e$). It is clear that the properties of the given crystal and the frequency range are of importance here.

The calculations presented below do not pretend to give an exhaustive description of various conceivable physical cases, even in the near-threshold energy region. The following relatively detailed analysis concerns cases which: (a) most clearly show the difference between the semiconductor–electrolyte and semiconductor–vacuum interfaces, (b) allow the calculations to be compared with experiment, and (c) illustrate the above-mentioned qualitatively different methods of description.

9.2. Calculation of the Photoemission Current at Semiconductor Electrodes

General Relationships of the Wave Approach

The following quantitative description of photoemission from semi-conductors in the near-threshold frequency region starts with the propagation of the probability wave described by the stationary single-particle Schrödinger equation. It is also assumed that (a) the contribution of processes proceeding without conservation of the transverse quasi-momentum component is negligible, and (b) the conduction band and the top of the valence band do not overlap and, moreover, all electron states are free in the conduction band and filled in the valence band.

The general expression (1.1) can be written in this case in the form

$$I = e \int j_x(E_1, \mathbf{p}_{\parallel}, \omega) \frac{2 \, d\mathbf{p}_{\parallel} \, dp_{1x}}{(2\pi\hbar)^3} \tag{9.2}$$

Equation (9.2) was derived from Eq. (1.1) by taking into account that the energy E_1 is unambiguously determined for the given quasi-moments $\mathbf{p}_{\parallel}$ and p_{1x} and that

$$p(E_1, \mathbf{p}_{\parallel}) \, dE_1 \, d\mathbf{p}_{\parallel} = \frac{2}{(2\pi\hbar)^3} \, d\mathbf{p}_{\parallel} \, dp_{1x}$$

The integration range in Eq. (9.2) is determined by the laws of energy and momentum conservation being given, as previously, by the condition

$$2m[E_1(\mathbf{p}_1) + \hbar\omega] - \mathbf{p}_{\parallel}^2 > 0 \tag{9.3}$$

Equation (1.12) can be utilized for calculating the partial flux of emitted electrons in the near-threshold energy range. The quantity $|\Lambda|^2$ retains here its approximate nondependence on E_f if the width of the energy range ΔE_f is smaller than the energy of the electron's motion in the region $x < 0$ and the resonance levels are absent. In any case, the corresponding condition for electrons in the crystal is satisfied if

$$\Delta E_f / \min\{E_g, |\chi|\} \ll 1 \tag{9.4}$$

where the denominator in Eq. (9.4) is the smallest of the quantities in brackets.

The inequality (9.3) is, in general, more rigorous than (1.11), which ensures constancy of $|\Lambda|^2$ in the case of emission from metals. Also, using any model of the surface barrier, the quantity $|\chi|$ is eliminated from Eq. (9.4), which reduces to

$$\Delta E_f / E_g \ll 1 \tag{9.4'}$$

Calculation of j_x requires, according to Eq. (1.12), knowledge of the potential profile $V(x)$ in Eq. (1.7), which describes the electron interactions outside the emitter.

Calculation of $V(x)$ is in this case a separate and generally rather complex problem. In the description of the long-range "tail" of the $V(x)$ potential in vacuum we can utilize the expression for the potential of image forces

$$V(x) = -\frac{e^2}{4x} \frac{1 - \varepsilon}{1 + \varepsilon}$$

where ε is the dielectric constant of the emitting semiconductor. In the case of emission into highly concentrated electrolytes, $V(x)$ can be assumed to be zero outside a narrow transition layer (Fig. 9.1), as in the case of emission from metals.

Frequency Dependence of the Photoemission Current

Consider now the photoemission current I at frequencies allowing only those electrons to be emitted which were initially localized close to the top of the valence band. Let it be assumed that in this case the effective mass approximation can be used; then $E_i = E_v - p_i^2/2m_v$, where $E_v < 0$ is the top of the valence band and $m_v > 0$ is the effective mass of the electrons in the valence band before excitation. Condition (9.3) can be now rewritten in the form of two inequalities:

$$\hbar\omega + E_v - \frac{p_\parallel^2}{2m}\left(1 + \frac{m}{m_v}\right) > \frac{p_{1x}^2}{2m_v}$$

$$0 < p_{1x}^2 < 2m_v(\hbar\omega + E_v) \tag{9.3'}$$

If $j_x(p_{1x}, p_\parallel)$ is known, Eq. (9.2) together with conditions (9.3') yield the solution of the problem.

Let $\omega_0 < \omega < \omega_0'$, i.e., direct transitions in the bulk resulting in photo-emission are forbidden. In vacuum and in the presence of significant image forces, $j_x = (\mathbf{p}_e^*/m)|\Lambda|^2$ [in full analogy with Eq. (1.16)], where $\mathbf{p}_e^*$ is a quantity obtained from $\mathbf{p}_e$ [cf. (1.16)] by replacing ε^{-1} by $(\varepsilon - 1)/(\varepsilon + 1)$, and $|\Lambda|^2$ is a constant independent of p_{1x} and $\mathbf{p}_{\|}$. Taking into account Eq. (9.3') and integrating Eq. (9.4) first with respect to $d\mathbf{p}_{\|}$ and then to dp_{1x}, we obtain

$$I = B(\hbar\omega - \hbar\omega_0)^{3/2} \qquad (9.5)$$

where

$$B \equiv \frac{4}{3}\frac{|\Lambda|^2}{\pi^2\hbar^3}\frac{p_e^*/m_v^{1/2}}{1 + m/m_v}$$

is independent of $\omega - \omega_0$. Equation (9.5) was first derived by Kane (1962).

The most important case here is photoemission in a polar medium where the image forces are eliminated. From Eqs. (1.16), (9.1), and (9.3') we have

$$I = B'(\hbar\omega - \hbar\omega_0)^2 \qquad (9.6)$$

where

$$B' = \frac{3}{64}\frac{|\Lambda'|^2}{|\Lambda|^2}\frac{m^{1/2}}{\mathbf{p}_e^*}B$$

and $|\Lambda'|^2$ is a constant, as before.

For frequencies $\omega > \omega_0'$, electrons with sufficient energy to escape from the semiconductor can be generated by the mechanism of direct transitions. Calculation of the photocurrent must then take into account that $\mathbf{p}_i = \mathbf{p}_f$. Assuming the parabolic dispersion law in the bonds

$$E_i = E_v - (1/2m_v)(p_{1x}^2 + \mathbf{p}_{\|}^2) \quad \text{and} \quad E_f = E_c + (1/2m_c)(p_{fx}^2 + \mathbf{p}_{\|}^2)$$

where E_c is the bottom of the valence band and $m_c > 0$ is the effective mass of the electron in this band, Eq. (9.1) can be rewritten as

$$E_g + \frac{p_{1x}^2 + \mathbf{p}_{\|}^2}{2m_c}\left(1 + \frac{m_c}{m_v}\right) - \hbar\omega = 0 \qquad (9.7)$$

The minimum momentum sufficient for the escape of the electron from the crystal when $\chi > 0$ is given by $\mathbf{p}_f = \{2m_c\chi)^{1/2}, 0\}$ and thus we obtain from Eq. (9.7)

$$\hbar\omega_0' = E_g + \chi\left(1 + \frac{m_c}{m_v}\right) = \hbar\omega_0 + \frac{m_c}{m_v}\chi$$

Equation (9.7) represents also an additional limitation of the range of permitted p_{1x} and $\mathbf{p}_{\|}$ values for the original electrons [in the momentum space the integration in Eq. (9.2) is not carried over the volume element determined by conditions (9.3') but over a certain surface determined by Eq. (9.7) and included in that volume].

Calculating, as previously, the magnitude of j_x and integrating, we obtain the following results: (a) for photoemission into vacuum (Kane, 1962):

$$I = C(\hbar\omega - \hbar\omega_0') \tag{9.8}$$

(b) for photoemission into a concentrated electrolyte solution (Gurevich, 1972)

$$I = C'(\hbar\omega - \hbar\omega_0')^{3/2} \tag{9.9}$$

where C and C' are constants independent of $\omega - \omega_0$ or of $\omega - \omega_0'$.

It should be mentioned that the effective mass approximation was used here only for simplicity. Equations (9.8) and (9.9) are obtained for any given dispersion law, although the relation between ω_0 and ω_0' is rather more complex in the general case.

Equation (9.8) was found to be valid in numerous experiments concerning photoemission at semiconductor–vacuum interfaces (cf., e.g., Philips, 1966). The "3/2 power law" [Eq. (9.9)] should apply for photoemission into electrolytes in all these cases when Eq. (9.8) is valid for the semiconductor–vacuum interface.

Thus, photoemission of electrons from semiconductors in concentrated electrolytes can be described (within the framework of the assumptions made) as follows. At light frequencies $\omega_0 < \omega < \omega_0'$ the process obeys Eq. (9.6). In the $\omega > \omega_0'$ range the emission current given by (9.6) is supplemented by an additional term of the type (9.9). The photocurrent due to the generation from the bulk is considerably higher than that resulting from the surface excitation mechanism at comparable values of $\omega - \omega_0$ and $\omega - \omega_0'$. Therefore, when $\omega > \omega_0$ Eq. (9.9) should apply predominantly outside a narrow transition region. Therefore Eqs. (9.6) and (9.9) may be combined (using the definitions of ω_0 and ω_0' as above) to give

$$I = \begin{cases} 0 & \text{for } \hbar\omega < E_g + \chi \\ B'(\hbar\omega - E_g - \chi)^2 & \text{for } E_g + \chi < \hbar\omega < E_g + \dfrac{m_v + m_c}{m_v}\chi \\ C'\left(\hbar\omega - E_g - \dfrac{m_v + m_c}{m_v}\chi\right)^{3/2} & \text{for } \hbar\omega > E_g + \dfrac{m_v + m_c}{m_v}\chi \end{cases} \tag{9.9a}$$

Note that $C' \gg B'$. For a more complex dispersion law in the semiconductor bands, $(m_v + m_c)/m_v$ in Eqs. (9.9a) is to be substituted by a certain factor also exceeding unity.†

† A new attempt to calculate photoemission current from semiconductor electrodes was made recently by Brodskii and Tsarevskii (1975b). Their results (in the only case when they are correct) identically transform to (9.9a).

Deviations from (9.9b) may be caused possibly also by violation of the condition $d_x/l \ll 1$.

The Effect of the Electrode Potential

The possible effect of the potential difference φ applied to the interface will now be considered. At sufficiently high electrolyte concentrations, the entire potential drop is virtually confined within the semiconductor (with the exception of highly doped semiconductors or of electrodes with a high density of surface states). This potential drop results in an additional deformation of the bands close to the surface (Myamlin and Pleskov, 1967). The width of a significant deformation equals the Debye length κ_{sc}^{-1} of the semiconductor [determined by the same equation as in the case of electrolyte solutions (p. 30) in which the term

$$\sum_i Z_i^2 n_i$$

is replaced by twice the value of the free carrier concentration in the bulk of the semiconductor and ε_0 by the dielectric constant of the semiconductor]. In the case of "intrinsic" germanium, for example, $\kappa_{sc}^{-1} = 10^{-4}$ cm. The effect of the band deformation described above can be neglected if $l \ll \kappa_{sc}^{-1}$.

The effect of potential on the energy characteristics of a metal and a semiconductor is shown schematically in Fig. 9.3 for the often encountered case when the entire potential drop is confined to the space-charge region of the semiconductor. [Under certain conditions (cf. Myamlin and Pleskov, 1967) a part of the potential drop can occur within the Helmholtz region of the double layer.] It can be seen from Fig. 9.3 that the additional potential applied to the semiconductor electrode does not change the values of the threshold constants ω_0 and ω_0'.

Thus, as opposed to the case of photoemission at the metal–electrolyte interface, the external potential applied to the semiconductor–electrolyte interface does not appreciably alter the laws of the photoemission step itself. (The possibility of indirect effects, consisting, for example, in changing the character of adsorption, or in affecting the number of carriers in bands close to the surface, is obviously not excluded.)

The basic physical causes of the difference in the behavior of externally polarized metal and semiconductor electrodes are as follows. First, the main factor which determines the change of photoemission under conditions of polarization is the change of the chemical potential (Fermi level) with respect to the surrounding medium. In the case of semiconductors, the main factor is not the change of chemical potential but of the position of the boundaries of the valence and conduction bands. Therefore, in spite of the change of chemical potential caused by the applied potential φ, the threshold frequencies

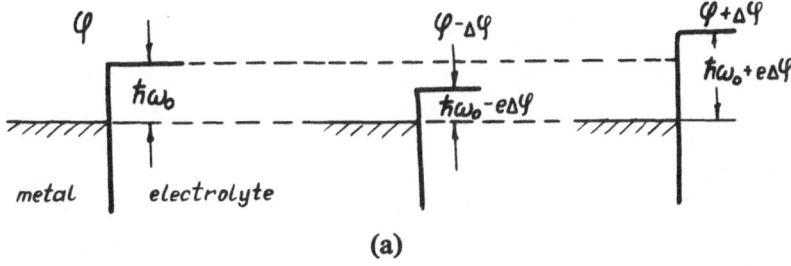

(a)

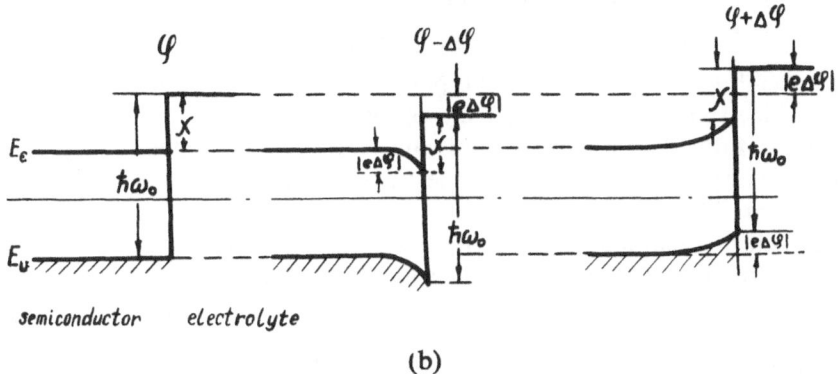

(b)

Fig. 9.3. Comparison of the effect of the applied potential difference on photoemission from metal (a) and semiconductor (b) electrode; $\Delta\varphi > 0$.

ω_0 and ω_0' remain the same. Second, the potential drop at the metal–electrolyte interface is confined to the ionic part of the double layer of width commensurate with atomic distances, whereas at the semiconductor-electrolyte interface the potential drop involves distances of the order of κ_{so}^{-1}, which normally considerably exceed atomic dimensions. Correspondingly, in the latter case, the distance over which the potential drop occurs always considerably exceeds the de Broglie wavelength of the electrons.

On the other hand, under certain conditions a part $\delta\varphi$ of the applied potential difference is localized within the Helmholtz layer. Accordingly, in this case the value of χ changes by $e(\delta\varphi)$ so that

$$\chi(\varphi) = \chi(0) + e(\delta\varphi) \tag{9.9b}$$

where $\chi(0)$ is the affinity at the electrode potential conditionally taken to be zero. Relations (9.9a) together with (9.9b) give the dependence of I on $\delta\varphi$. It should be borne in mind that, generally speaking, $\delta\varphi$ depends on φ and therefore the dependence of I on φ may be rather complex.

The Effect of Surface States

The simplest model of photoemission from surface states will now be considered. These states are assumed to form a surface band (cf. Fig. 9.1). The extent to which the surface zone is filled with electrons is obviously determined by the position of its upper, E_u, and lower, E_l, boundaries with respect to the level of the chemical potentials of the semiconductor, μ ($E_l < E_u < 0$ for the chosen zero energy level). In particular, if $E_u < \mu$, the surface band is completely filled. The calculation of the photoemission current from surface states, $I^{(s)}$, using the parabolic dispersion law close to the boundary of the surface zone, results in (Gurevich, 1972)

$$I^{(s)} \propto (\hbar\omega - \hbar\omega_0^{(s)})^{3/2} \tag{9.10}$$

where $\hbar\omega_0^{(s)} = -E_u$ determines the photoelectric threshold from the surface zone. Since the threshold frequency $\omega_0^{(s)}$ is less than ω_0, the photoemission current is fully determined by emission from surface states. Therefore, in spite of its relatively low density, the current can, in principle, be observed experimentally.

It will be assumed now that $E_l - \mu > kT$ and the surface zone is filled to a very small extent. At frequencies $\hbar\omega > E_l$ we obtain

$$I^{(s)} \propto \exp[(\mu - E_l)/kT] \tag{9.11}$$

Since μ changes linearly with φ, the dependence of $I^{(s)}$ on φ is exponential in the range considered. Over a wider range $E_l < \mu < E_u$ the dispersion law for the surface band must be known.

Calculation of the Photocurrent in Terms of Diffusion Phenomena

The motions of photoelectrons in the bulk crystal will now be briefly considered from the diffusional viewpoint. The most interesting objects of such studies are semiconductors exhibiting a negative electron affinity, whose properties have been successfully described in these terms. Since the bottom of the conduction zone is higher in such semiconductors than the energy level of a stationary electron in vacuum, not only "hot" but also thermalized electrons can escape from the crystal. Let it be assumed that the thermalization of excited electrons is very fast and the corresponding length is small compared with η^{-1} and L, where η^{-1} is the absorption coefficient (of the same order of magnitude as l_ω) and L is the diffusion length. The motion of electrons in the crystal can then be approximately described by the diffusion equation

$$-\mathscr{D}\frac{d^2 n_e}{dx^2} + \frac{n_e}{\tau} = \frac{J}{\hbar\omega}\eta e^{-\eta x} \tag{9.12}$$

where n_e is the concentration of thermalized electrons, $\mathscr{D}$ is the diffusion coefficient, τ is their lifetime in the conduction band, $L = (\mathscr{D}\tau)^{1/2}$, and J is the energy of the light flux absorbed by the electrode. It can be seen from the right-hand side of Eq. (9.12) that the light absorption in the bulk is completely due to the photoexcitation of electrons. Solution of this equation, fully analogous to that carried out in Section 2.2, with suitable boundary conditions, allows the density of the electron flux from the bulk crystal to the interface to be determined. Such equations (but of a more specific nature) were utilized by James and Moll (1969).

Denoting by p the probability that an electron thermalized close to the interface will escape from the semiconductor, the quantum yield is expressed by

$$Y = \frac{p}{1 + (\eta L)^{-1}} \tag{9.13}$$

The Y^{-1} vs. η^{-1} dependence should be linear with intercepts $\eta^{-1} = L$ on the abscissa and p on the ordinate. In the above approximation, the dependence of Y on ω is fully determined by $\omega(\eta)$.

The simple considerations described above have led to good agreement with experiment in the frequency range immediately bordering on the fundamental absorption band.

It can be seen from Eq. (9.13) that the specific properties of the interface and, in particular, the dependence of the quantum yield on the electrode potential are determined in this approximation by the phenomenological parameter p. Correspondingly, Eq. (9.13) can be applied both to the semiconductor–vacuum and semiconductor–electrolyte interfaces, so long as the form of p remains undefined.

The question of the applicability of the diffusion, or wave mechanism of bulk excited electron transfer at higher light frequencies, remains as yet open. The condition which, in terms of the wave approach, allows photoemission to occur at $\chi < 0$ is $p_{tx} > 0$. The threshold frequency is determined by $\hbar\omega_0 = E_g$, and when $\hbar\omega > E_g$ we obtain from Eq. (9.7) the expression for the photocurrent (Gurevich, 1974)

$$I = C''(\hbar\omega - E_g) \tag{9.14}$$

where C'' is a constant independent of the sum $(\hbar\omega - E_g)$.

9.3. Photoemission from Semiconductor Electrodes : Experimental

Photoelectron emission from a semiconductor into an electrolyte was first observed by Krotova and Pleskov (1973). Subsequently the phenomenon was studied by Meyer (1973) and Boikova *et al.* (1975).

Photoemission experiments carried out with semiconductor electrodes require separation of the photoemission and photoconductive effects, i.e., excitation of electrons from the valence to the conduction band, without their escape from the crystal. The photoconductive effect is observed at quantum energies exceeding the width of the forbidden band of the semiconductor. The bulk photoexcitation (cf. Section 9.1) always accompanies photoemission. However, the photoconductive effect is by itself also a source of the photocurrent, even if photoelectron emission is not induced. In particular, an increase of the minority carrier concentration accelerates electrode reactions in which these carriers take part, thus resulting in a photocurrent (Myamlin and Pleskov, 1967). With increasing light frequency, a threshold is reached above which the photocurrent due to the photoconductive effect is supplemented by the photoemission current.

Experimentally, the photoemission current can be separated by measurements carried out in the presence and in the absence of acceptors in solution. The latter measurements yield only the photoconductive current, which requires no acceptors. Thus the difference between the two currents is the photoemission current (assuming that the acceptors do not enter into any electrochemical "dark" reactions by themselves).

The electrode should consist of a doped semiconductor of the p-type,

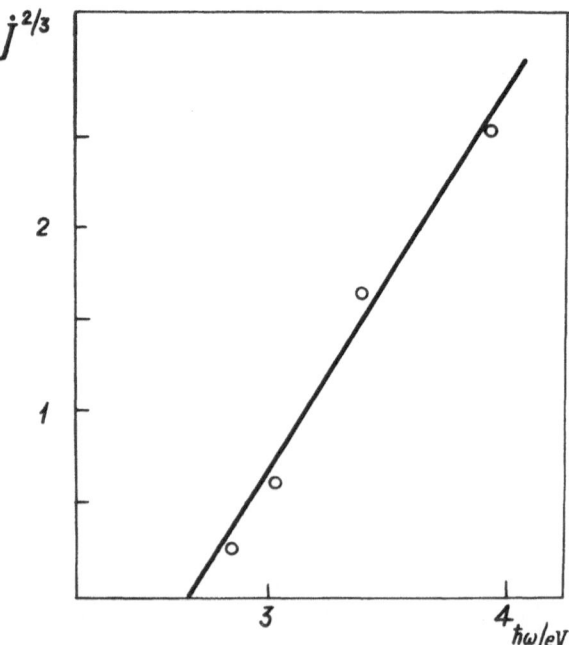

Fig. 9.4. The 3/2 power law and the determination of the photoelectric threshold at a germanium electrode (Krotova and Pleskov, 1973). p-type germanium, 0.003 Ω-cm; 0.2 M KCl solution saturated with N_2O; potential -1.3 V.

since cathodic polarization does not result, in this case, in any considerable "dark" currents. Otherwise, the photoemission measurements are carried out in the same way on semiconductor as on metal electrodes.

The spectral characteristics of the photocurrent observed at a constant potential at a germanium electrode are shown in Fig. 9.4 as a plot of $j^{2/3}$ vs. $\hbar\omega$. The plot obeys, within the experimental error, the 3/2 power law (9.9), valid in the case of bulk excitation of photoelectrons. Larger photocurrents, which might be ascribed to the surface mechanism of photoexcitation, were not observed in this experiment (although they appear apparently in other cases). Extrapolation of the curve obtained to $j = 0$ results in the work function, $\hbar\omega_0' = 2.7$ eV (at -1.3 V). As in the case of metal electrodes, the work function measured at the semiconductor–solution interface is considerably lower than in vacuum (the work function of germanium in vacuum is $w_v = 4.8$ eV) owing to the interaction of the electron with the solvent and to the effect of cathodic polarization.

The dependence of photocurrent on potential at a germanium electrode is more complex (Fig. 9.5). The very existence of this dependence indicates that a considerable part of the externally applied potential difference must involve the ionic part of the double layer. (The potential drop in the space-charge region in the semiconductor does not affect the photoemission current;

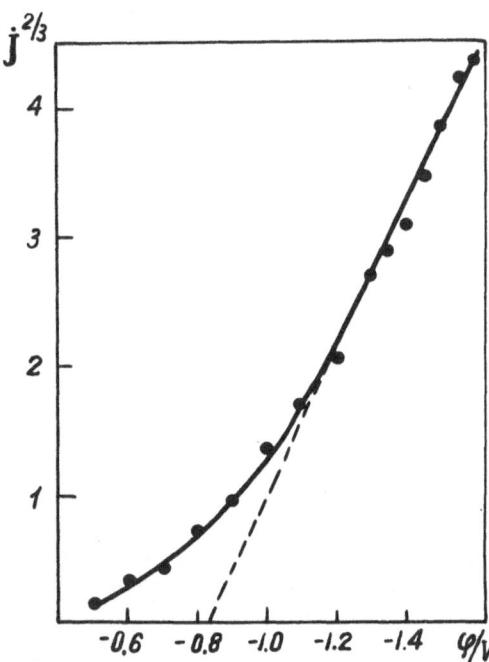

Fig. 9.5. The voltametric characteristics of the photocurrent at a germanium electrode (Boikova *et al.*, 1976). 0.1 *M* LiClO₄ solution saturated with N₂O.

cf. Section 9.2.) At high negative potentials, a linear $j^{2/3}$ vs. φ section exists. This must be connected with the externally applied potential difference being virtually confined to the ionic layer. At less negative potentials, the photocurrent varies as a weaker function of potential. This can be interpreted in terms of a considerable part of the externally applied potential dropping within the space-charge region and thus not being "useful" with respect to the energy of the photoemission process. Boikova *et al.* (1976) calculated the potential distribution between the space-charge region in the semiconductor and the ionic part of the double layer. Their results are in good agreement with the independent evaluations of Myamlin and Pleskov (1967). Similar behavior is observed for photoelectron emission from gallium arsenide.

The angular dependences of the photoemission current at a germanium electrode illuminated by polarized light is shown in Fig. 9.6b (Meyer, 1973), together with similar curves for the anodic photocurrent which is entirely due to the photoconductive effect in the semiconductor (see Fig. 9.6a). As much as the anodic curve practically coincides with the absorption curves, the photoemission current exhibits considerable divergences from the absorption curve obtained for light polarized in the plane of incidence. In

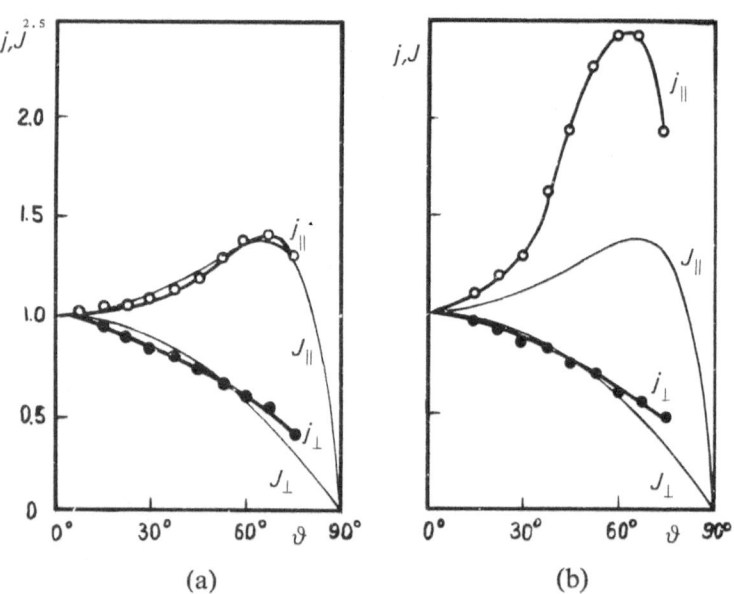

Fig. 9.6. The photocurrent j and the absorbed energy of the light flux J as functions of the angle of incidence of the illuminating light (Meyer, 1973). Light polarized parallel ($\parallel$) to the plane of incidence and perpendicular ($\perp$) to that plane. 1 M HClO$_4$ solution at potentials: (a) 0.3 V, (b) -0.3 V (NHE); quantum energy 4.1 eV.

this respect, the germanium electrode resembles metallic ones, for which similar divergences have been observed (cf. Figs. 4.10 and 4.11).

Determination of the photoemission energy characteristics is exemplified by the behavior of a p-type germanium electrode. In Fig. 9.7 (cf. Fig. 9.1) the electron energy levels in metal and semiconductor electrodes are compared (at -1.3 V; cf. Fig. 9.4). The electrochemical potentials are the same at the same electrode potential. From the value of the work function of the metal in solution w_{ms} [cf. Eq. (4.5)] we can immediately obtain the electrochemical potential of the semiconductor: $\mu = -2.1$ eV. (The potential energy of a delocalized electron in the electrolyte solution is again chosen as the

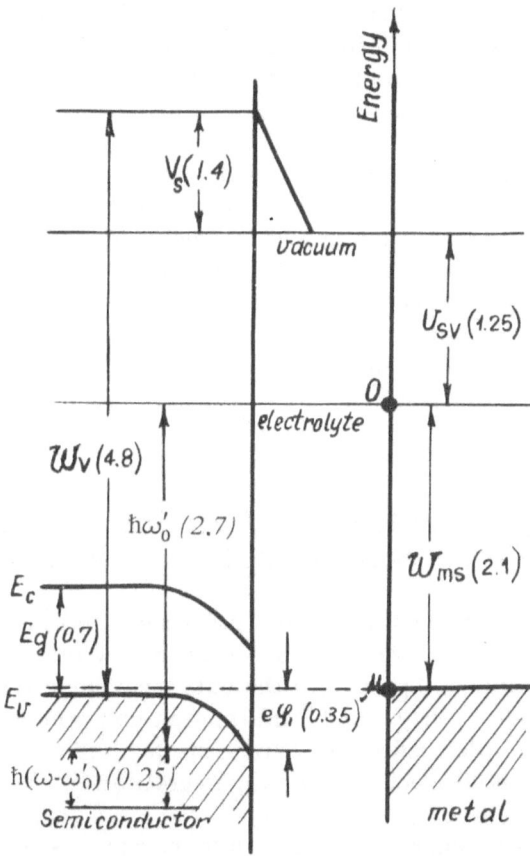

Fig. 9.7. Energetics of photoemission from metal and p-type germanium electrodes in solutions at -1.3 V. The absolute energy values (eV) are shown in brackets (Boikova *et al.*, 1976).

arbitrary zero.) The electrochemical potential of a strongly doped p-type semiconductor can be practically identified, without any serious error, with the top of the valence band, E_v. The deformation of the boundaries of the energy bands of the semiconductor $|e\varphi_1|$ at this potential (also determined from photoemission measurements, Boikova *et al.*, 1976) is 0.35 eV. From the known energy level of the top of the valence band near the surface, and the values of the work function of the metal in solution (2.1 eV) and of the measured work function $\hbar\omega_0'$ of the semiconductor (2.7 eV), the "depth" of the photoemitting level in the valence band can be calculated as the difference between the work functions corresponding to the bulk and surface excitation, $\hbar(\omega - \omega_0') = 0.25$ eV (represented in Fig. 9.7 by the dashed arrow).

Finally, the Volta potential between germanium and the solution can be calculated from the equilibrium work function of germanium in vacuum ($w_v = 4.8$ eV), the work function of the metal in solution (2.1 eV), and the difference between the energy levels of a delocalized electron in aqueous solution and in vacuum ($U_{sv} = 1.25$ eV). From the calculated value, $V_s = 1.4$ eV, the Volta potential can be determined for any chosen electrode potential. It should be mentioned that the measurement of Volta potentials between an electrode and solution is, in general, a very difficult experimental problem. The photoemission method provides a relatively easy way of determining this quantity.

The study of photoemission properties of semiconductor electrodes has only relatively recently been started. It can be hoped that the photoemission method will open new perspectives for the investigation of the semiconductor-electrolyte interface.

9.4. Photoelectron Emission from Solutions into the Vapor Phase. Schematic Aspects of the Process

Studies of photoemission from a polar solution in equilibrium with its vapor complement and verify to a certain extent the results described in the previous sections. Apart from the data concerning the structure of the liquid–gas interface, they supply information concerning the energy states of solvated and delocalized electrons in the liquid, the mechanism of retardation of "hot" electrons, and the interaction of photoactive ions with light.

Photoelectron emission from solutions was discovered by Stoletov (1888b) who studied aqueous solutions of fuchsin. The phenomenon was interpreted in terms of a photoactive layer formed at the interface by the molecules of the dye, from which electrons can be relatively easily ejected.

Subsequently, photoemission was observed both from aqueous (Allen, 1925; Hughes and Du Bridge, 1932) and nonaqueous solutions including alkali metals in liquid ammonia (Kraus, 1921; Haesing, 1940; Teal, 1947). A systematic quantitative study of the basic laws of electron photoemission from solutions into the gas phase was initiated in the late sixties by Delahay and co-workers.

Thanks to modern techniques of photoemission measurement, a large body of experimental material has been accumulated which serves as a basis for the development of a theory of the phenomenon.

The description will be limited here to the case of photoemission from solutions containing, as the electron source, localized (solvated) electrons. This case resembles in many ways photoelectron emission from solutions containing other electron donors (cf. Baron *et al.*, 1970; Delahay *et al.*, 1970; Nemec and Delahay, 1972; Nemec *et al.*, 1972; Ballard, 1972; Ballard and Griffiths, 1960, 1971).

It should be mentioned here that the solvated electron can exist for periods of time sufficient for measurements only in a limited number of solutions (e.g., liquid ammonia, hexamethylphosphotriamide). The lifetime of a hydrated electron in water is of the order of milliseconds and therefore photoemission from aqueous solutions involves mainly anion emitters, such as I^-, $Fe(CN)_6^{4-}$, $Mo(CN)_8^{4-}$, and others.

Photoemission from solutions might be treated as direct experimental evidence for the possibility of photoionization of solvated electrons, i.e., of the optical transition from a solvated state into a "conduction band." However, it must be stressed that the formation of delocalized electrons upon illumination of solvated-electron solutions is possible not only as a result of direct photoionization, but also by a two-step mechanism: an optical excitation of the electron from the equilibrium (e.g., $1s$ level) to a higher level still in the bound state, followed by its subsequent autoionization (Baron *et al.*, 1970).

According to Delahay's model (1971), photoemission of electrons from the liquid into the gas phase can be represented by the following steps:

(a) Formation of free, or delocalized, electrons in solution upon absorption of light by photoactive substances ("emitters") as a result of their photoionization.

(b) Propagation of the delocalized electrons in solution toward the liquid–vapor interface.

(c) Escape of the delocalized electron into the vapor phase. This process requires the electron to overcome the potential barrier (including the surface potential) at the liquid–vapor interface.

(d) Motion of the electrons (due to the electromagnetic field) in the gas phase toward the anode, resulting in the observed photocurrent. A fraction of the emitted electrons, the magnitude of which depends on the applied field and vapor pressure, returns in the solution.

Thus, photoelectron emission from solutions is a multistep process and the observed photocurrent depends on each of the separate stages.

It follows from the above scheme that a threshold energy of the light quantum should exist in order to make photoemission possible. For a discussion of the relation of this threshold energy to other characteristic features of the process, it is useful to consider again Fig. 4.9. The measured work function w_{sv} differs from the solvation energy A_s, i.e., from the thermodynamic equilibrium work of transfer of the electron from its bound state in solution to the vapor phase. In fact, photoionization of the "emitter" results, in agreement with the Franck–Condon principle, in a nonequilibrium state of the solution; i.e., the solution molecules retain the same orientation around the un-ionized emitter as they had before the escape of the electron. Therefore, the photoionization energy A_{op} is not equal to the difference of the energy levels of an electron in the bound state and the bottom of the conduction band in solution, but exceeds this difference by the reorientation energy E_s. This "excess" energy which must be expended to achieve photoionization is subsequently transferred to the solvent, being thus "useless" for the following photoemission step.

It follows from the above discussion and from Fig. 4.9 that in the case where the emitter is a solvated electron

$$w_{sv} = A_s + E_s, \qquad A_{op} = A_s + E_s - U_{sv} \qquad (9.15)$$

The quantities which appear in Eq. (9.15) can, in principle, be measured experimentally. The equilibrium work function A_s can be determined by thermal emission measurements from solutions containing solvated electrons (the phenomenon discovered by Baron *et al.*, 1970, using solutions of sodium in hexamethylphosphotriamide). The nonequilibrium (or, as it is called by Delahay, "vertical") work function w_{sv} can, in general, be obtained from experimental curves of the dependence of the photocurrent j_s on the quantum energy $\hbar\omega$ by extrapolation to $j_s = 0$. A coordinate system suitable for such an extrapolation can be chosen if the photoemission law is known. Finally, the photoionization energy can be determined from absorption spectra of solutions of solvated electrons if the light absorption is due to photoionization (or if the part corresponding to photoionization can be separated from the total spectrum); see, for example, Lugo and Delahay (1972); Nemec (1973); Brodskii and Tsarevskii (1973).

The theory of photoelectron emission from solutions must take into account all stages of the overall process described above.

9.5. Principles of the Theory of Photoelectron Emission from Solutions

A complete theory of the phenomenon does not exist as yet. Therefore we are bound to limit the scope of this section only to a qualitative description of facets of the photoemission process.

The foundations for theoretical considerations of photoemission from solutions were laid by Delahay (1971) who suggested that the measured photocurrent j_s can be described by

$$j_s = ea\mathscr{P}L \tag{9.16}$$

where $\mathscr{P}$ is the rate of generation per unit volume of delocalized electrons with sufficient energy to escape from the solution, L is a parameter (expressed in units of length) which characterizes the thickness of the solution layer supplying emitted electrons, and a is a dimensionless coefficient accounting for the return of the emitted electrons to the solution. Equation (9.16) should also include a factor corresponding to the reflection of photoexcited electrons in solution from the solution–gas interface. This factor will be further assumed to be included in a.

The dependence of $\mathscr{P}$ on $\hbar\omega$ was calculated by Delahay, following equations for the cross section of photoionization of a hydrogenlike atom. The number of delocalized electrons formed per unit time is proportional to the photoionization cross section σ_{ph} of the interaction of a photon with an electron in its ground state which results in the latter's transition into a continuous spectral state in vacuum, and therefore the cross section σ_{ph} can be calculated using the following approximate expressions (Bethe and Salpeter, 1957):

$$\sigma_{ph} \propto \begin{cases} (A_{op}/\hbar\omega)^{8/3} & \text{for } A_{op} \leqslant \hbar\omega < 2A_{op} \\ (A_{op}/\hbar\omega)^3 & \text{for } \hbar\omega > 2A_{op} \end{cases} \tag{9.17}$$

(It should be mentioned that A_{op} represents the energy of an electron transition from a lower state of the emitter, with the most probable configuration of the solvent, to a delocalized state with zero kinetic energy and the same solvent configuration.) The energy range $\hbar\omega < A_{op}$ excluded from Eq. (9.17) is of no interest here, since low-energy electrons cannot escape from the solvent. It should also be mentioned that the general form of the dependence of the photoionization cross section on quantum energy $\hbar\omega$ of the type $\sigma_{ph} \propto (\hbar\omega)^{-b}$ (where $b \simeq 3$) is retained when $\hbar\omega > A_{op}$ for effective potential wells of various types; they need not necessarily be hydrogenlike ones, as assumed above. In particular, the localized electron can be located in a rectangular potential well (Kajiwara *et al.*, 1972). Thus the generation rate can be expressed by $\mathscr{P} \propto c_e(A_{op}/\hbar\omega)^b J$, where c_e is the bulk concentration of emitters, b is equal to 3 or $\frac{8}{3}$, and J is the energy of the absorbed light flux.

With these assumptions, the spectral characteristics $\mathscr{P}$ should resemble the short-wave end of the absorption spectrum of the given emitter solution, since both the process of light absorption and of photoemission are connected with the electron transition from the ground (localized) state into the band. For solutions of alkali metals in certain amines, the short-wave end of the spectrum is in fact described by a function of the type $\sim(\hbar\omega)^{-8/3}$. A more complex dependence is observed for aqueous ferrocyanide solutions. A detailed analysis of absorption spectra of emitting solutions can be found in the papers by Lugo and Delahay (1972) and by Copeland *et al.* (1970).

The dependence of the parameter L on the quantum energy or, rather, on the energy of the delocalized electron will now be considered. The mean distance covered by the electron in the liquid is determined by processes of its resolvation and capture by scavengers (e.g., impurities in the solution). The distance L characterizes the thickness of the solution layer from which electrons can escape into the gas phase. The electrons formed by light absorption in deeper-lying layers will be "lost" on their way to the surface and will not contribute to the photocurrent. The quantity L is closely connected with the retardation distance of electrons with a given initial energy in a condensed medium and resembles in a physical sense the electron thermalization length for electrons emitted from a metal into solution. The latter distance is in turn connected with the mean solvation length x_0 (cf. Section 5.3). The length L can, in principle, be determined from comparison of Eq. (9.16) with experimental data (see below).

The necessary data concerning electron scattering in the vapor can be obtained from the photoemission experiments at the metal–gas interface. The Thomson formula

$$j_s = I_s \frac{\mathscr{F}/p}{u + \mathscr{F}/p} \tag{9.18}$$

is normally used for this purpose, where j_s is the measured photocurrent, I_s is the emission current, $\mathscr{F}$ is the electric field between the emitter and collector, p is the gas pressure, and u is a parameter determined by the properties of the gas phase. The emission current I_s can be calculated from plots of j_s^{-1} vs. $\mathscr{F}$ (at constant p).

Delahay (1971) did not consider the effect of the reflection of photoelectrons from the interface, thus making an implicit assumption that the transfer coefficient through the surface barrier is unity for electrons with sufficient energy to escape from the solution. This transfer coefficient was discussed by Brodskii and Tsarevskii (1974). According to their scheme, photoionization results in a probability wave extending around each emitter. The attenuating wave propagates to the interface and becomes partly reflected. The photocurrent corresponds to the transferred wave. The attenuation in the bulk is formally taken into account by the "optical" potential with an imaginary component; this is equivalent to the introduction of the phenomenological length L mentioned above.

Brodskii and Tsarevskii assumed further that (a) the surface potential barrier has no long-range "tail" due to image forces, and (b) the final energies of all emitted electrons are relatively small (in particular, lower than the height of the surface potential barrier). Then, for light incident normal to the interface (experimental conditions of Delahay and his co-workers), the photoemission current should have the following dependence on light frequency (Brodskii and Tsarevskii, 1974):

$$I_s \propto (\omega - \omega_0)^{5/2} \tag{9.19}$$

where ω_0 is the threshold frequency for photoemission from solution.

Whether or not the threshold conditions are satisfied depends on the choice of the frequency range of the illumination. The effect of image forces (cf. Section 9.2) on an electron emitted from a dielectric with a dielectric constant ε into vacuum is described by the potential $V(x) = [(1 - \varepsilon)/(1 + \varepsilon)](e^2/4x)$. Designating $\varepsilon_{\text{eff}} \equiv (1 + \varepsilon)/(1 - \varepsilon)$, it can easily be seen that the potential $V(x) = e^2/4x\varepsilon_{\text{eff}}$ has the same form as that for the case of emission from a metal in a dielectric with the dielectric constant ε_{eff}. In the latter case, however, the parameter $\gamma \equiv \hbar(\omega - \omega_0)/E_1$ plays the most important part; in the present case $\gamma = \hbar(\omega - \omega_0)\varepsilon_{\text{eff}}^2/33.5$ [$\hbar(\omega - \omega_0)$ is measured in eV]. If $\gamma \gg 1$, the image forces can be neglected; if $\gamma \ll 1$ the long-range "tail" is of considerable importance. The problem thus reduces to the proper choice of ε. For its maximum (static) value $\varepsilon_0 \gg 1$, $\varepsilon_{\text{eff}} \simeq 1$, $\gamma \ll 1$, so that the image forces cannot be neglected. The minimum logically admissible ε_0 value (with the corresponding maximum ε_{eff}) corresponds here to the visible frequency range. Since $\varepsilon \simeq n^2$, where n is the refractive index, we obtain from Table 4.1, for example, for liquid ammonia, $\varepsilon = 1.75$ and $\varepsilon_{\text{eff}} = 3.7$. The dielectric constant of other solvents used in photoemission experiments is even higher and correspondingly ε_{eff} even lower. Substitution of ε_{eff} in the expression for γ shows that even in this "optimum" case, the image forces cannot be neglected.

Thus, Eq. (9.19) can hardly be used for the correct treatment of experimental data. Moreover, as was already mentioned in Section 1.4, the transfer coefficient of the surface barrier is close to unity in the presence of image forces. Therefore, the above estimate can serve as supporting evidence for the phenomenological assumption implicitly made in Delahay's theory that the transfer coefficient is equal to unity.

The approach of Brodskii and Tsarevskii supplemented by inclusion of the effects of image forces should lead to a relation $I_s \propto (\omega - \omega_0)^2$, similar to Eq. (1.20) describing photoelectron emission from metal into a vacuum (or vapor).

9.6. Photoelectron Emission from Solutions. Basic Experimental Results

The experimental apparatus used by Delahay and co-workers (details of its construction are described in the original papers of Baron et al., 1969, and Delahay et al., 1970) includes the optical part (light source, monochromator, focusing lenses) and the electrical measuring circuit. The wavelength range used depends on the nature of the solution studied. Alkali metal solutions in hexamethylphosphotriamide, for example, are illuminated with light at wavelengths in the 1200–3300 Å range. Aqueous solutions of anion emitters (e.g., ferrocyanides) require ultraviolet light (2500–2000 Å). The quantum yield referred to the incident light flux does not usually exceed 10^{-4} to 10^{-5} electron/photon (the low yield is explained by the depth of electron escape L, being much lower than the thickness of the light absorbing layer). Therefore the measured photocurrents are extremely small (10^{-11} to 10^{-8} Å), and special measures must be undertaken to prevent current leaks (in particular, a shielding electrode). The anode is shaped in the form of a flat helix oriented parallel to the surface of the solution, thus creating a uniform field without hindrance to the incident light.

Experimental data for a solution of sodium in hexamethylphosphotriamide plotted according to Eq. (9.18) are shown in Fig. 9.8. Intercepts on the ordinate obtained by extrapolation of linear sections observed at high $\mathscr{F}$ values serve to determine the photoemission current.

It should be remembered that the field applied between the anode and solution affects, in the general case, not only the electron drift, but also the height of the potential barrier at the solution–vapor interface (Shottky effect, mentioned in the Introduction). The latter effect was found to be negligible in the case of photoemission from solutions (Delahay, 1971); it must, however, be considered in the case of thermal emission (Baron et al., 1970).

The spectral characteristics of the photoemission current from a sodium solution in hexamethylphosphotriamide are shown in Fig. 9.9. The curve exhibits two maxima (which are also found in the absorption spectrum of this solution). The first maximum (long wavelength one) is ascribed by Baron et al. (1971) to the emission of electrons formed by photoionization of solvated electrons themselves, and the second (short wavelength one), by photoionization of a more stable complex of solvated electron with Na^+ ion (apparently a bielectron complex $Na^+ \cdots e_2^{2-}$, described recently by

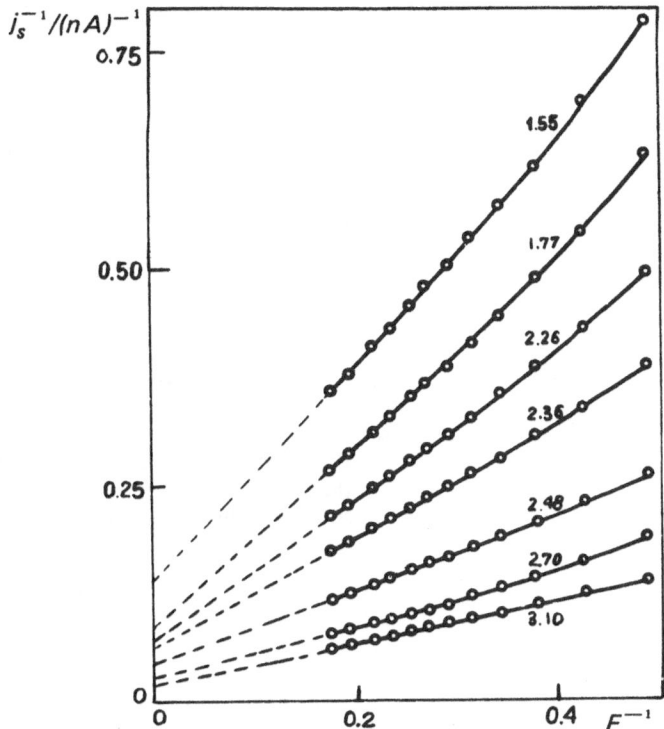

Fig. 9.8. Determination of the photoelectron emission current from solution by extrapolation of photocurrents measured at various field intensities (Delahay, 1971). 0.114 M solution of sodium in hexamethylphosphotriamide; $t = 5.6°C$; vapor pressure 200 mm Hg. Quantum energies (eV) shown in the figure.

Alpatova and Grishina, 1973, and Alpatova *et al.*, 1973a, b). Further discussion concerns the first maximum.

The photocurrent increases with sodium concentration (and consequently of solvated electrons) in solution in a rather complex fashion. The possible reasons for this are discussed below.

The effective work function w_{vs} was determined by Delahay (1971) from $I_s(\hbar\omega)^b - \hbar\omega$ by means of plots (cf. Section 9.5) which correspond to the expression for photocurrent: $I_s \propto Jc_e(A_{op}/\hbar\omega)^b L$. Here J is the absorbed power of the light flux. The parameter b equals 3 or $\frac{8}{3}$ (see above). The results of such a treatment are shown in Fig. 9.10. (For sodium solutions in hexamethylphosphotriamide the section of the I_s vs. $\hbar\omega$ curve in the region of the long-wavelength maximum is considered; cf. Figs. 9.9 and

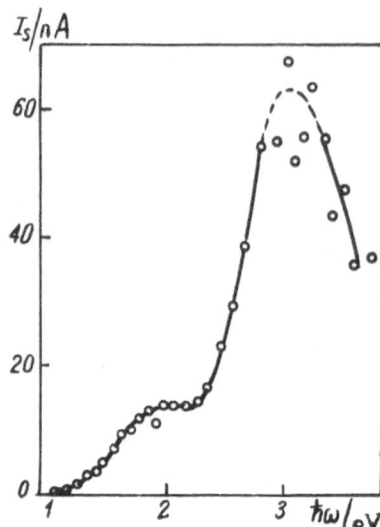

Fig. 9.9. The spectral characteristics of the photoelectron emission from 0.114 M sodium solution in hexamethylphosphotriamide (Baron *et al.*, 1971).

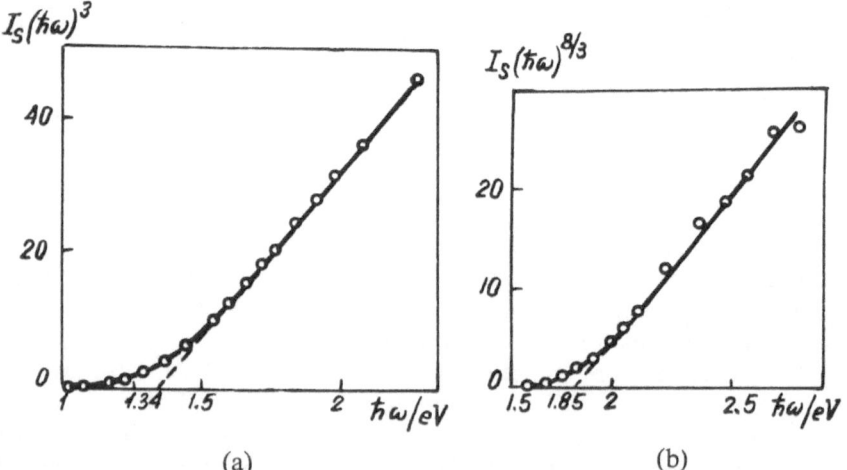

(a)   (b)

Fig. 9.10. Determination of the photoelectric threshold for solutions (Delahay, 1971). (a) 0.114 M sodium solution in hexamethylphosphotriamide; (b) 1% sodium solution in liquid ammonia (calculated from Haesing's data, 1940).

9.10a.) In the chosen coordinate system, straight lines are indeed obtained, which, extrapolated to the intercept on the axis of abscissae, yield the threshold energy which is equal, according to Delahay, to the work function w_{sv}.

For sodium solutions in hexamethylphosphotriamide and liquid ammonia, $w_{sv} = 1.34$ and 1.85 eV, respectively; for aqueous solutions of $K_4Fe(CN)_6$ it is 5.95 eV. The author connects deviations from linearity observed in the immediate vicinity of the threshold with the equilibrium energy distribution of the initial "emitter" configurations.

The same experimental data were plotted by Brodskii and Tsarevskii (1975a) as $I_s^{0.4}$ vs. $\hbar\omega$. Their results for sodium solutions in liquid ammonia are shown in Fig. 9.11. In agreement with the theory of the latter authors, the "near-threshold" part of the I_s vs. $\hbar\omega$ curve lends itself to this treatment. The values of the threshold energies thus obtained are lower than those of Delahay, being equal for the three systems mentioned above to 0.9, 1.53, and 5.46 eV, respectively.

Finally, Gremmo and Randles (1974) obtained plots of the value of w_{sv} for a sodium solution in hexamethylphosphotriamide, giving 1.05 eV by means of an approximate extrapolation of $\log I_s$ vs. $\hbar\omega$.

It is too early yet to judge the validity of the desired work function values. Apart from the incomplete theory, the accuracy of photocurrent measurements for solution–vapor systems is low, so that experimental data themselves cannot indicate the choice, for example, between the $\frac{3}{8} = 0.385$ or the 0.4 power law. It should be mentioned, however, that the treatment

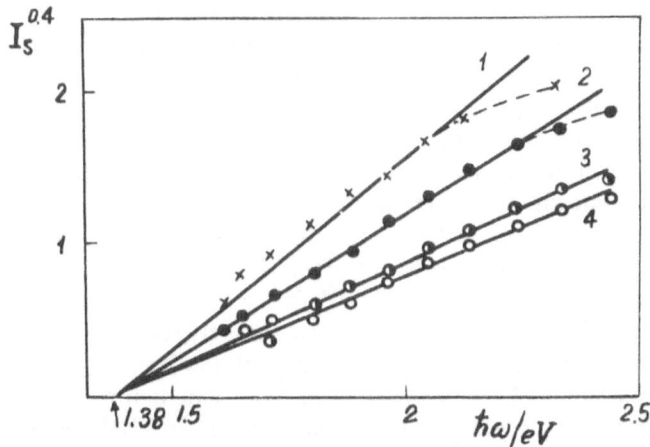

Fig. 9.11. Determination of the photoelectric threshold for solutions (Brodskii and Tsarevskii, 1975a). Potassium solutions in liquid ammonia. Molar concentration: (1) 0.51; (2) 0.13; (3) 0.32; (4) 0.08 (experimental data of Aulich *et al.*, 1973).

of experimental results by Brodskii and Tsarevskii (1974, 1975a) seems inconsistent. The latter authors used the threshold approximation which assumes low final energies of emitted electrons, in particular, in comparison with the height of the potential barrier at the interface. However, it can be seen from Fig. 9.11, for example, that this energy $\hbar\omega - w_{sv}$ reaches 1 eV, i.e., the same magnitude as the height of the barrier U_{sv} at the ammonia interface determined by the same authors. For hexamethylphosphotriamide, the barrier height should be lower than w_{sv}, i.e., less than 0.9 eV. Thus, the first condition of applicability of the threshold approximation, viz. low energy of emitted electrons in comparison with the barrier height, does not hold in the energy range used for their graphical interpretation. This puts in doubt results obtained by Brodskii and Tsarevskii, at least within the accuracy claimed.

Delahay (1971) tried to determine the dependence of L on the energy of delocalized electrons in the solution. On the basis of the observed slopes of $I_s(\hbar\omega)^b$ vs. $\hbar\omega$ curves, he concluded that a linear dependence exists between L and the electron energy (see also Aulich et al., 1973, and Delahay, 1973). This conclusion and the estimated value of $L = 40$ Å are in good agreement with the dependence of x_0 on the electron energy, found in photoemission experiments at the metal–solution interface (see Section 5.3). The agreement confirms the validity of concepts concerning the character of motion of delocalized electrons in solution described in Section 5.3.

We shall now briefly describe thermoelectric emission from solutions of solvated electrons. As was already mentioned, the equilibrium work function can, in principle, be determined from the temperature dependence of the thermoelectron emission current. Baron et al. (1970) quote the value of 1.0 eV for a sodium solution in hexamethylphosphotriamide. According to Gremmo and Randles (1974), however, the work function varies from 0.95 eV in a 0.05 M sodium solution to 0.74 eV in 10^{-3} M solution. It should be mentioned that the values obtained in this fashion are somewhat undefined. In fact, the complex character of the photoemission and thermoemission currents on sodium concentration indicates the existence of specifically adsorbed sodium at the interface. The surface potential due to this adsorption depends on temperature and thus it must affect the temperature dependence of the thermoemission current. Therefore the value obtained from the temperature dependence of the thermoemission current can differ from the work function. The possibility also exists of the equilibrium between the localized and delocalized electrons in solution being disturbed during thermoemission. The equilibrium concentration of delocalized electrons, for example, in hexamethylphosphotriamide, is very low (cf. Section 10.3) and therefore the thermoemission rate can be controlled by the transition of solvated electrons into the delocalized state (i.e., in the band).

Specific Problems of Photoelectrochemical Phenomena

The increasing interest in photoelectrochemical and especially in photo-emission phenomena is reflected in the large number of theoretical and experimental papers recently published concerning various aspects of the overall photoprocess, which may include, as one of the steps, the photo-emission of electrons. The "Becquerel effect," as mentioned previously, consists, in the general case, of a whole complex of separate photoprocesses. Some of them, including photoemission, can form the basis of new methods of investigation of electrode kinetics and of the structure of electrode-electrolyte interfaces. Moreover, the qualitative and quantitative charac-teristics of photoelectron emission into electrolytes can be compared with the properties of the photoemission process into vacuum and dielectrics. Such a comparison and consideration of several other effects (of a non-emissive nature) of light on electrochemical systems are the subject of the present chapter.

10.1. Currents of Photoelectrochemical Reactions

According to Heyrovsky (1967a, b), the photoeffect at the metal–electrolyte interface is due to the decomposition of a charge-transfer complex, formed by the surface of the metal electrode with solvent or solute molecules. It follows from the preceding material that the concept of photo-decomposition of charge-transfer complexes cannot pertain to the photo-electrochemical properties of a large number of systems investigated at potentials more negative than the photoemission threshold. In this potential range, the photoprocess is primarily due to photoemission. However, for more positive potentials, prohibitive energetically to photoemission, this interpretation seems to be valid. Photocurrents observed under these con-ditions are usually anodic.

Anodic photocurrents are observed in solutions of organic compounds (e.g., oxalic, chloroacetic, and tartaric acids, and their salts) containing usually a carboxyl group and an electronegative substituent at the neigh-boring carbon atom. Upon adsorption at the electrode surface, the electronegative atoms form a chemical bond with the metal. It can be assumed that this bond decomposes under illumination, the electrons are transferred to the electrode, and the oxidized organic molecule passes into the solution. If the adsorption is fast and the solution layer in the immediate vicinity of the electrode does not become depleted in the adsorbate, a stationary photo-current can be observed. Photocurrents, supposedly due to charge-transfer complexes, are (as opposed to photoemission currents) insensitive to light polarization, as can be seen, for example, from the experiments carried out by Korshunov *et al.* (1969) in oxalic acid solutions; cf. Fig. 4.12. Another difference arises from the great sensitivity of the currents in question to the properties of the adsorbate (Heyrovsky, 1967b). It should be remembered that photoemission currents are virtually insensitive to the chemical nature of the acceptor (as long as the secondary processes involving electron capture products, [eA], remain insignificant).

The mechanism of processes initiated by absorption of light by sub-stances adsorbed at the electrode surface can be very complex, as illustrated by a probable path leading to the appearance of the anodic photocurrent at a mercury electrode in an oxalate solution. According to Barker and Concialini (1973b), the photocurrent does not depend directly on the elec-trode potential but is determined by the "surface excess" of the oxalate at the interface, which in turn is potential dependent. Upon absorption of light the adsorbed oxalate anion transfers its electron to the electrode

$$(COO)_2^{2-} \xrightarrow{(\hbar\omega)} \begin{array}{c} COO^- \\ | \\ COO \end{array} + e^-(M)$$

The radical anion formed remains adsorbed and decomposes according to the reaction

$$\begin{array}{c} COO^- \\ | \\ COO \end{array} \rightarrow CO_2 + CO_2^{\cdot-}$$

followed by electro-oxidation of the $CO_2^{\cdot-}$ radical ion

$$CO_2^{\cdot-} \rightarrow CO_2 + e^-(M)$$

Thus, the photodecomposition of the surface complex itself is followed by subsequent electrochemical and chemical steps.

An interesting hypothesis concerning the mechanism of photocurrents has recently been suggested by Gerischer (1973) and by Gerischer *et al.* (1972). The excitation of the electron in the metal caused by light absorption is accompanied by the simultaneous formation of an unfilled level, i.e., of a hole (as in the case of light absorption in semiconductors); cf. Fig. 10.1.

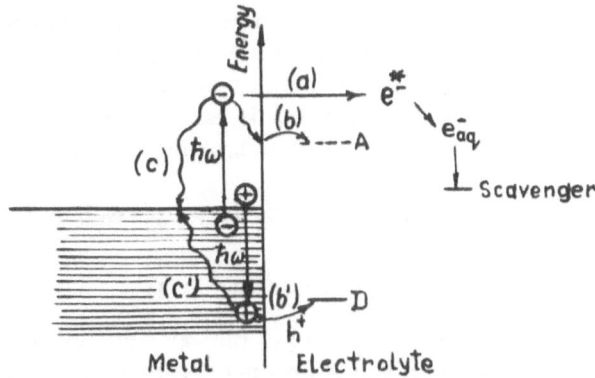

Fig. 10.1. Photoprocess involving excited electrons and holes in a metal (Gerischer, 1973). (a) Photoemission; (b, b') transition of the excited electron or hole to the acceptor (A) or donor (D) level at the electrode surface; (c, c') an excited electron and a hole lose energy and recombine.

The maximum "depth" of the hole level in the metal is $\hbar\omega$ below the Fermi level. The basic assumption here is the existence, for a finite period of time, of such a free level before it becomes filled with electrons from higher levels. A fraction of the holes (and of simultaneously formed excited electrons) lose energy in interactions with the lattice and recombine (processes c and c' in Fig. 10.1). The lifetime of a hole is assumed to be sufficient for a considerable fraction of the holes to reach the electrode surface, by analogy with the situation in semiconductors (cf. Myamlin and Pleskov, 1967). (It should be stressed that the propagation of holes in crystals is a conventional concept describing the actual propagation of an electron in the opposite direction.) The hole can then be transferred to the donor level at the electrode surface (process b' in Fig. 10.1). If this process is irreversible, i.e., the oxidized donor particle does not undergo further reduction at the electrode, an anodic current is observed. In a similar way, an excited electron can be captured by an adsorbed acceptor molecule (process b in Fig. 10.1) without passing through a delocalized state in solution (corresponding to photoemission — path a in Fig. 10.1).

Although the mechanism presented above is called by the authors "photoemission of holes," in reality emission of holes into solution does not occur. In fact, in solution, the final state is localized, i.e., by definition (cf. p. 11) this process (not emission but rather photoexcitation of the metal) is analogous to that proposed by Berg *et al.* (1967) in their theory of the "hot" electron (cf. Section 0.1).

The anodic photocurrents which can be observed at gold and platinum electrodes in sulfuric acid solutions can serve as examples of "hole photo-currents." They appear in the potential range in which, according to Gerischer *et al.* (1972), oxide films are absent but the photoconductive effect might take place. Photoelectrochemically active oxide films are formed on platinum and gold at more positive potentials; cf. Vinnikov *et al.* (1973, 1974).

Water molecules at the electrode surface are assumed to act as electron donors:

$$H_2O + h^+ \rightarrow OH^. + H^+$$

where h^+ designates a hole. Hydroxyl radicals can further react with ad-sorbed OH^- ions to form a stable product, hydrogen peroxide:

$$OH^. + OH^- \rightarrow H_2O_2 + e^-(M)$$

The latter can, in general, become oxidized at the electrode

$$H_2O_2 \rightarrow O_2 + 2H^+ + 2e^-(M)$$

The dependence of the photocurrent on the quantum energy and on the electrode potential (Fig. 10.2) qualitatively resembles the behavior of the

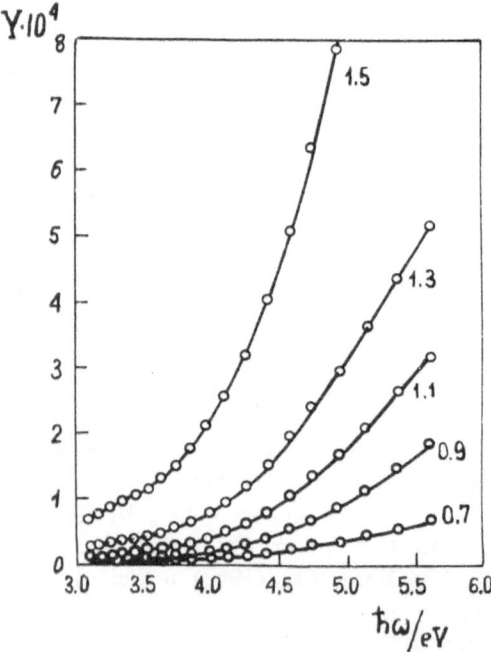

Fig. 10.2. The quantum yield of the anodic photocurrent at a gold electrode as a function of quantum energy at various potentials (Gerischer, 1973). 0.5 M H_2SO_4. Potentials (in volts) shown in the figure.

cathodic (emission) photocurrent. The angular characteristics of the photo-current (Fig. 4.11) led Gerischer *et al.* (1972) to suggest the bulk mechanism of electron (and hole) photoexcitation in metals. The maximum quantum yield is about 10^{-3} electron/photon.

Of course, the mechanism discussed above remains hypothetical and requires further experimental evidence. In particular, a quantitative verification of the consequences resulting from the suggested model is needed.

Another attempt at a theoretical interpretation of photocurrents in electrochemical systems was made by Matthews and Khan (1975). Without going into details of this paper, we shall comment briefly on some aspects of it. First, the cathodic photocurrents discussed by the authors are really due to the direct transfer of electrons to acceptors in the electrolyte and are thus not really photoemission currents, contrary to the authors' statement. Second, within the framework of the actual mechanism discussed, the central problem is the calculation of the probability of the electron transition from the metal directly into a localized state on the acceptor. The authors, however, replace this calculation by that of a tunneling probability across a surface barrier model formed by adsorbed water molecules.

The laws of photocurrents of the types discussed in this section, as well as the problem of substances which cause their appearance, are relatively little known. Further investigations can be expected to supply important data concerning, in particular, the energy of surface levels corresponding to adsorbed particles.

10.2. Currents of the Pulse Warm-Up of the Electrode

Berg and co-workers were the first to observe photocurrents due to the "warming up" of the electrode and the adjacent solution layer caused by absorption of the light. These warm-up currents were ascribed to changes of the differential capacity with temperature. The use of lasers causes the local temperature to reach 100–200°C. However, even low-powered light sources such as pulse flash lamps cause changes of the surface temperature sufficient to result in measurable warm-up currents.

The differential capacity of an electrode in dilute solutions is practically equal to the capacity of the diffuse layer, described in the case of a 1:1 electrolyte by the following expression (Delahay, 1965):

$$C_{\mathrm{d}} = F\left(\frac{\varepsilon_0 c_{\mathrm{el}}}{2\pi RT}\right)^{1/2} \cosh \frac{F\psi'}{2RT} \qquad (10.1)$$

where ε_0 and c_{el} are the static dielectric constant and concentration of the electrolyte, respectively. According to Eq. (10.1) the capacity decreases with increasing temperature. Under galvanostatic conditions, the change of capacity results in a shift of potential, i.e., in the appearance of a photo-potential; under potentiostatic conditions, it results in the appearance of a current in the external circuit (cf. the equivalent circuit in Fig. 3.1). In both cases, the sign of the effect depends on the sign of the electrode charge. On this basis, Korshunov *et al.* (1968a) and Zolotovitskii *et al.* (1971) suggested a method of determination of the point of zero charge of metals. According to these authors, the photoeffect caused by the pulse warm-up should change its sign at the potential of zero charge.

The experiments were carried out using long-wavelength light which did not induce photoemission. The temperature increase was estimated by the authors to be within 1°C. The method is, in fact, a variant of the well-known general method of the temperature jump, based on measurements of relaxation of a chosen parameter after a sharp change of temperature of the system (cf. Caldin, 1964). According to Zolotovitskii *et al.* (1971) the method gives satisfactory results for a series of metals. However, a more thorough analysis of this technique demonstrates certain difficulties in its application.

First, the measurements were carried out in relatively concentrated solutions (0.1–1 *M*) for which the temperature dependence of the differential capacity differs from Eq. (10.1). (This problem was not discussed at all in the papers cited.) In fact, the measured capacity is virtually equal in such solutions (especially at potentials far removed from the point of zero charge) to the capacity of the compact layer. The latter can both decrease (as in the case of the diffuse-layer capacity) or increase with increasing temperature; moreover, its temperature dependence differs for various metals. The compact-layer capacity becomes temperature independent at a certain potential which depends on the nature of the metal and the composition of the solution. According to Grahame (1957), the temperature coefficient of the differential capacity of mercury in 0.1 *M* solutions of alkali metal fluorides becomes zero at -0.2 V (i.e., far from the point of zero charge). If the capacity is temperature independent, the appearance of a pulse warm-up current cannot be expected. In other words, this current should be zero in concentrated electrolytes also at the potential at which the capacity is temperature independent. However, this was not observed by Zolotovitskii *et al.* (1971).

Second, the results may be affected by the thermoelectromotive force which in galvanic cells can reach the same order of magnitude (10^{-3} V-deg^{-1}, Temkin, 1953) as the measured warm-up potential. Therefore, the results described must be treated with great caution. Moreover, recent results of Barker and Cloke (1974) show that the change of sign of the photoeffect observed under conditions of pulse warm-up does not coincide with the potential of zero charge. The potential of zero photoeffect is 0.1–0.15 V more negative than the potential of zero charge of mercury in 0.01–0.2 M solutions of potassium, sodium, and lithium halides, as illustrated in Fig. 10.3. (It is clear from Fig. 10.3 that, as opposed to the results of Zolotovitskii *et al.*, 1971, the warm-up current, j_h, curve tends to a second intersection with the potential axis in the region of positive potentials, i.e., exactly in the region where the temperature coefficient of the differential capacity is close to zero.)

It must be concluded that the heating of the electrode due to absorption of light brings about a whole series of effects. Thus, Barker and Cloke (1974) consider not only the decrease of the diffuse-layer capacity but also the possibility of the change of orientation of adsorbed solvent molecules (which affects the Galvani potential at the interface), as well as thermal diffusion effects in solution.

The pulse warm-up effect thus requires further experimental and theoretical investigation, both for ideally polarizable and completely nonpolarizable electrodes.

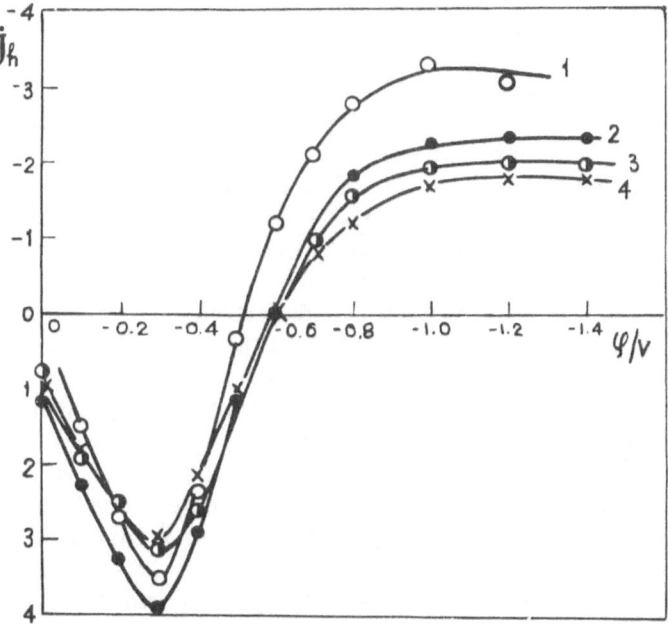

Fig. 10.3.   Pulse warm-up currents at the mercury electrode (Barker and Cloke, 1974). 0.05 M solutions of HCl (1), KCl (2), NaCl (3), and LiCl (4).

10.3. Cathodic Generation of Solvated Electrons. Is the Solvated Electron an Intermediate in Cathodic Reactions?

Photoemission methods supplied the first possibility of reasonably accurate determination of a series of energy characteristics for electron transfer across the metal–solution and solution–vapor interfaces (cf. Sections 4.3 and 9.6). The data obtained can be utilized in various calculations connected, for example, with the analysis of energies of "dry" and solvated electrons, with problems of the relation between the Galvani and Volta potentials, with considerations of homogeneous and electrode reactions, etc.

Here we shall briefly discuss the application of photoemission studies to the development of general ideas concerning the mechanism of electrode reactions. The successes of radiation chemistry, crowned by the discovery of the solvated electron in a series of solvents (including water), inspired some authors to assume that transfer of electrons from the cathode into solution is a prior step in a number of cathodic reactions. This transfer was regarded as being followed by formation of solvated electrons which then react with substances present in solution (Antropov, 1971; Pyle and Roberts, 1968; Kenney and Walker, 1971; Walker, 1967b).

A detailed discussion of this hypothesis is beyond the scope of this book, but a thorough examination of this question can be found in the papers by Conway (1972) and Conway and MacKinnon (1970). Formation of hydrated electrons near the electrode surface was recently investigated in detail by several authors. A review of results obtained in this new field was published by Krishtalik and Alpatova (1976). We shall very briefly consider these aspects of the cathodic generation of solvated electrons which can be naturally related to photoemission problems.

The cathodic generation of solvated electrons can proceed by various paths: (a) via an intermediate deposition at the electrode of an alkali metal (during discharge of a cation of an indifferent electrolyte) and its subsequent chemical dissolution; (b) via a direct transfer of metal electrons into traps formed by corresponding fluctuations of the orientation of solvent dipoles (so called "electrochemical dissolution" of electrons); (c) via the thermal-emission mechanism, i.e., the transfer of thermally excited electrons from the metal into solution where they exist for a finite time in a delocalized state before undergoing solvation.

The direct cathodic generation of electrons can be experimentally observed in hexamethylphosphotriamide, liquid ammonia, and in certain amines and other solvents which do not decompose at the electrode at very negative potentials. In particular, noticeable generation can be observed in hexamethylphosphotriamide at -2.7 to -2.9 V with respect to the normal hydrogen electrode in the same solution. The solution near the electrode acquires a deep blue color characteristic of solvated electrons in this solvent.

The mechanism and kinetics of the cathodic electron generation are extremely sensitive to the state of the electrode surface. A great majority of metals are strongly passivated in hexamethylphosphotriamide at high negative potentials. Under these conditions the generation current obeys the Tafel equation with a slope of 60 mV (Fig. 10.4, curve 1). A systematic investigation of the process carried out at various metals and in several electrolytes led Krishtalik *et al.* to the conclusion that under the conditions employed, electrons are generated by thermal emission. In fact, the work function w_{ms} evaluated according to the Richardson–Sommerfeld equation for the thermal-emission current $I_T = A_0 T^2 e^{-w_{ms}/kT}$, where A_0 is the Sommerfeld constant [cf. Eq. (1.18)], virtually coincides with the w_{ms} value obtained from the analysis of photoemission data for a metal in hexamethylphosphotriamide (cf. Table 4.1).

The Tafel slope at active surfaces is 120 mV (Fig. 10.4, curve 2). In this case the mechanism of generation consists in "electrochemical dissolution" of electrons.

The value of the work function measured using the photoemission method for the transfer of electrons from a metal into a delocalized state in hexamethylphosphotriamide was used by Krishtalik and Alpatova (1976)

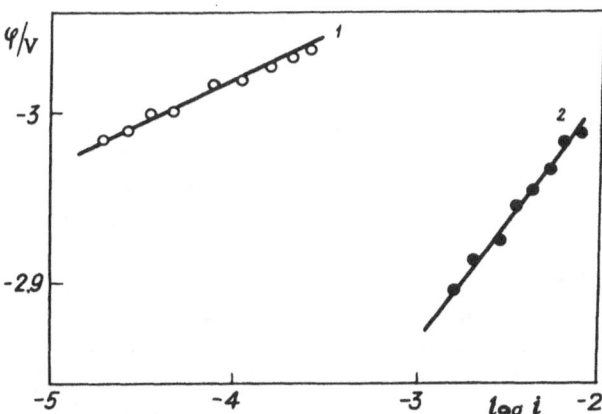

Fig. 10.4. Cathodic generation of solvated electrons (Krishtalik and Alpatova, 1975). Copper electrode with passivated (1) and active (2) surface. 0.32 *M* LiCl in hexamethylphosphotriamide, 5°C. Potential (V) measured with respect to the saturated (aqueous) calomel electrode.

to determine the energy difference between a delocalized and solvated electron in this solvent. Its value, -0.7 eV, was used for evaluation of the ratio of equilibrium concentrations of electrons in delocalized and solvated states, which was found to be $\sim 10^{-11}$.

Cathodic hydrogen evolution in hexamethylphosphotriamide proceeds at overpotentials lower by 1.5–2 V than the cathodic generation of electrons. Therefore, there is no doubt that the hydrogen evolution does not involve solvated electrons as intermediates (cf. Conway, 1972). The two processes, hydrogen evolution and electron generation proceeding independently of each other, have thus been clearly observed in hexamethylphosphotriamide.

The probability of thermal emission into aqueous solutions was estimated by Brodskii and Frumkin (1970), again on the basis of the "photoemission" work function. The probability was found to be very low, indicating a negligible contribution of the "thermal emission" mechanism to cathodic currents in aqueous solutions, even at very negative potentials. The probability of the usually accepted mechanism (i.e., of direct electron transfer to solvated ions or molecules) can be a few orders of magnitude higher at these potentials.

Thus, the numerical evaluations based on photoemission studies show little possibility that the concept of the intermediate formation of solvated electrons in cathodic reactions proceeding in aqueous solutions has any real basis.

10.4. Photoelectrochemical Effects Due to Surface Plasmons

One of the most interesting and widely discussed multiparticle effects in metals is the collective oscillation of the electron plasma in the bulk. Quasi-particles corresponding to these collective oscillations are called "plasmons" (Pines, 1963; Ziman, 1960, 1972). The plasma oscillations are due to the coulombic interactions of the electron system with the positively charged ionic lattice. Their frequency ω_p is given by $\varepsilon_{met}(\omega_p) = 0$, where $\varepsilon_{met}(\omega)$ is the frequency-dependent dielectric constant of the metal. For the simplest model of a metal $\varepsilon_{met} = 1 - (4\pi e^2 n_e/m\omega^2)u$ (cf. Section 1.1) and $\omega_p = (4\pi e^2 n_e/m)^{1/2}$. In the general case, the quantities $\varepsilon_{met}(\omega)$ and ω_p are complex numbers. The imaginary part of ω_p describes the attenuation in time of plasma oscillations.

Similarly, collective oscillations of the plasma type can arise at the boundary of the metal with a medium having different bulk properties. The corresponding quasi-particles are called "surface plasmons." Their frequency

ω_s is given by $\varepsilon_{met}(\omega_s) + \varepsilon_{med}(\omega_s) = 0$, where $\varepsilon_{med}(\omega)$ is the dielectric constant of the medium surrounding the metal (Ritchie, 1957; Stern and Ferrell, 1960). At the metal/vacuum interface, when $\varepsilon_{met}(\omega) = 1 - (\omega_p/\omega)^2$ and $\varepsilon_{med} = 1$, we have $\omega_s = \omega_p/\sqrt{2}$.

The problems connected with the properties and mechanism of excitation and decomposition of surface plasmons have attracted considerable attention in recent years in connection with photoemission studies. In particular, Ritchie and Wilems (1969) and Crowell and Ritchie (1970) demonstrated the possibility of surface plasmon excitation by illumination of the metal surface.

Subsequently, Endriz and Spicer (1971) showed that optically excited plasmons may decompose, transferring their energy to a single electron. If the decomposition energy exceeds the work function, the electron can be emitted from the metal. Thus, a photoemissive effect can arise due to a rather original photoexcitation mechanism.

The energies $\hbar\omega_p$ and $\hbar\omega_s$ corresponding to bulk and surface plasmons, respectively, are for a majority of metals of the order of 10 eV, i.e., they belong to the extrathreshold region. Silver is an exception, the energy of a surface plasmon at the silver–vacuum interface being approximately 3.6 eV. Moreover, the surface of silver in vacuum can be relatively easily maintained in a sufficiently clean state, making this metal extremely interesting for studies of surface plasmons. The excitation of surface plasmons at thin silver films illuminated in vacuum has been observed experimentally (Hoffmann and Steinmann, 1968). However, the plasmon energy is in this case lower than the work function in vacuum, so that photoemission is not possible. Decrease of the work function by means of adsorption of, for example, cesium or barium "spoils" the surface with respect to the possibility of excitation of surface plasmons.

Meanwhile, the work function of a metal in solution can be considerably lower than in vacuum (cf. Section 4.4), and the surface of a silver electrode in contact with solution can be made sufficiently clean and stable by suitable choice of the electrolyte composition. Therefore, an electrochemical cell with a silver photocathode can be conveniently used for investigating electron photoemission induced by surface plasmons. Such a study was recently made by Sass *et al.* (1974). The dependence of the quantum yield Y of photoelectron emission from a silver electrode on the light energy is shown in Fig. 10.5. A 2000 Å thick silver film served as the photocathode. The measurements were carried out in 0.5 M H_2SO_4 solutions (hydrogen ions served as acceptors). At energy values corresponding to the excitation of surface plasmons, the quantum yield curve exhibits a distinct peak never observed for metals previously examined. The position of the peak is virtually independent of potential; its height increases with increasing negative potential faster

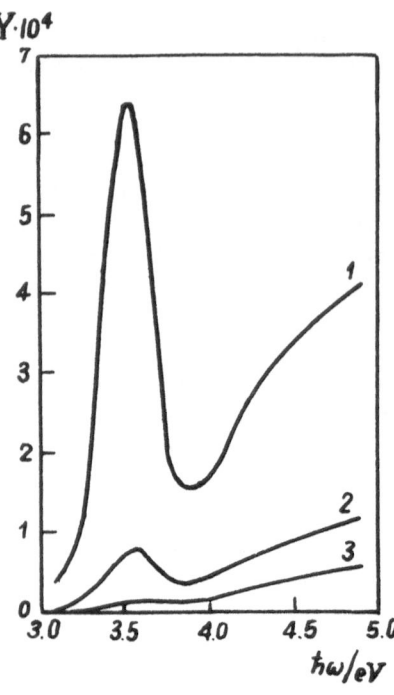

Fig. 10.5. Contribution of surface plasmons to photoemission (Sass *et al.*, 1974). Quantum yield of the photocurrent at a silver electrode as a function of the quantum energy in 0.5 M H_2SO_4 solution. Potential (vs. normal hydrogen electrode), volts: -0.2 (1); 0 (2); $+0.2$ (3).

than the quantum yield observed in other parts of the spectrum. The results obtained, together with the additional measurements of the reflection coefficient of the silver electrode, indicate that the photoprocess does indeed take place according to the mechanism described above.

Quite interesting results were also obtained in studies of anodic currents (Sass *et al.*, 1974) at silver electrodes (cf. Section 10.1). The presence of a peak resembling that observed for cathodic polarization (Fig. 10.5) indicates that surface plasmons also play an important role in this case. It can be assumed that the plasmon energy is transferred to the electron, resulting in photoexcitation. However, the electron energy at anodic potentials is insufficient to overcome the high work function, and the electron cannot then escape from the metal. The hole created by the photoexcited electron can, however, be captured at sufficiently high anodic potentials by a corresponding level created on the silver surface by solution components. The resulting electron transfer to the metal is observed as the photocurrent. The mechanism described for the anodic photocurrent is thus similar to that for "hole emission" discussed in Section 10.1 (Gerischer *et al.*, 1972), but is not quite adequate in our opinion. When the surface plasmons participate in electron photoexcitation, the probability of finding a hole near the surface can be considerably higher than in the "usual" mechanism of bulk excitation. Therefore, comparison of anodic currents observed at various metals would be of considerable interest. It should be mentioned that, in many cases, adsorption results in a decrease of the energy of surface plasmons (Gadzuk, 1970). Therefore, anodic currents observed under conditions of adsorption at other metals can, in certain cases, have the same origin as anodic photocurrents observed at silver electrodes.

10.5. Photoemission into Various Media

Our interest has been focused until now on photoelectron emission into electrolyte solutions. An additional review of certain aspects of this process in various media allows a deeper insight to be gained into a series of specific features of photoemission.

We shall first reconsider the fact that the threshold potential (cf. Section 4.3) of photoelectron emission at metal electrode–solution interfaces does not depend on the nature of the metal. It would seem that this is really a unique property, characteristic of this type of interface only. However, a certain analogy can be found with photoemission at the metal–vacuum interface. It will be recalled that the work function in vacuum, w_{mv}, is usually defined as the work of electron transfer from uncharged metal to infinity in vacuum. According to the simplest model, w_{mv} consists of the work done against the metal–electron "bond energy" (chemical potential μ of the electron) and of an electric component connected with the surface potential drop χ_m:

$$w_{mv} = -\mu + e\chi_m \qquad (10.2)$$

Now, if two metals, I and II, are brought in a contact equilibrium, a potential difference V_{12} arises (Fig. 10.6a):

$$eV_{12} = \mu_1 - e\chi_m^{(1)} - \mu_2 + e\chi_m^{(2)} \qquad (10.3)$$

where indices 1 and 2 correspond to metals I and II, respectively. The work of transfer of an electron from metal I to infinity is no longer equal to the work function w_{mv} but is given by $-(\mu_1 - e\chi_m^{(1)} - e\psi_1)$ (where ψ_1 is the outer potential of metal I) and has the same value as for metal II. For a charged surface, the work function w_{mv} equals [cf. Eq. (10.2)] the work of transfer of an electron from a metal to a distance (estimated as 10^{-4} to 10^{-5} cm), where the intensity of the field created by the surface charge has fallen effectively to zero. This is the work function measured by photoemission experiments at the metal–vacuum interface. In fact, as follows from Fig.

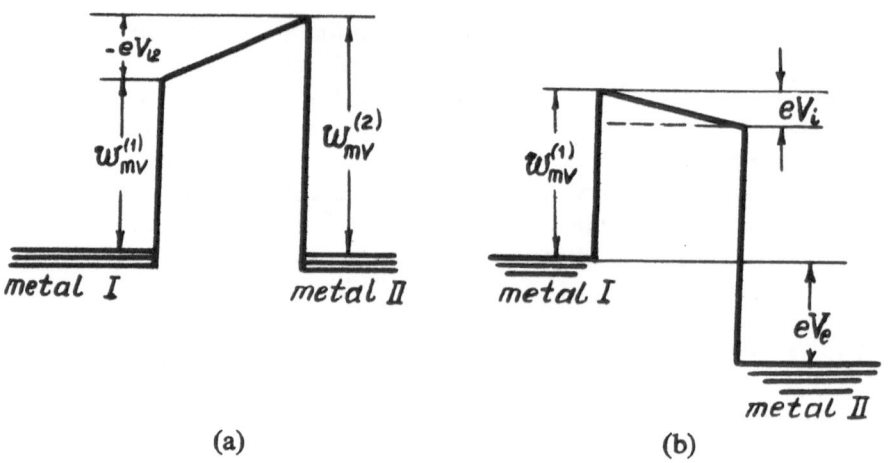

Fig. 10.6. Diagram illustrating the appearance of a contact potential between two metals in vacuum; (a) $V_e = 0$; (b) $V_e > 0$.

10.6a, under equilibrium conditions a contact potential difference exists between the emitting photocathode (I) and collecting anode (II). If $w_{mv}^{(1)} < w_{mv}^{(2)}$, then, according to Eq. (10.3), $\psi_1 < \psi_2$, i.e., the metal with a lower work function is positively charged. Upon cathode illumination the collector is reached only by electrons whose energy in vacuum exceeds the contact potential difference. All electrons emitted can reach the collector if an external potential V_e is applied ($V_1 = V_e + V_{12}$) which accelerates the drift of electrons to the collector (i.e., $V_1 < 0$). Then (cf. Fig. 10.6b) the work function in vacuum is indeed equal to the height of the barrier, $w_{mv}^{(1)}$, and is different for various metals. The total work of transferring an electron from the emitter to the collector differs, however, from the work function itself by the value of the inner potential difference $V_1 = V_e + V_{12}$ and is hence independent of the nature of the emitting metal. If the inner potential difference could be confined to a thin layer transparent to electrons (dashed line in Fig. 10.6), the electron work function would be independent of the nature of the metal. This is the case at the metal–electrolyte interface, where the Volta potential is indeed confined to the electric double layer of atomic dimensions.

We shall consider now the dependence of photoemission current at the metal–vacuum interface on the outer potential difference. This dependence is shown in Fig. 10.7 for two pairs of metals. In the first case, the contact potential between the cathode and anode, $V_{12}^{(1)}$, is positive (curve 1), and the second case, $V_{12}^{(2)}$, is negative (curve 2). The limiting photoemission

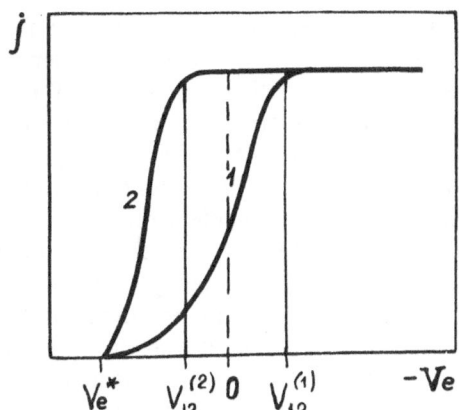

Fig. 10.7. Photoemission currents at
the metal–vacuum interface as a function of the outer potential difference.
(1) $V_{12} > 0$; (2) $V_{12} < 0$.

current (electrons are not returned to the emitter) is reached when $V_e + V_{12} = 0$. The potential V_e^* at which the photocurrent disappears (threshold potential) is independent of V_{12} and consequently of the nature of the emitting metal. In fact, the emitter potential measured with respect to the collector should satisfy the requirement that the inner potential difference $V_1 = V_e + V_{12}$ be sufficient for retardation of all emitted electrons. (This is the basis of the method of the retarding field, widely used in the studies of energy distribution of emitted electrons.) Since the maximum electron energy equals $\hbar(\omega - \omega_0)$, the condition for disappearance of the photocurrent caused by the retarding field is given by $\hbar(\omega - \omega_0) = V_e^* + V_{12}$. On the other hand, $eV_{12} = \hbar(\omega_0^{(2)} - \omega_0)$, where $\omega_0^{(2)}$ is the work function of the collector. Therefore, $eV_e^* = \hbar(\omega - \omega_0^{(2)})$. The value of V_e^* thus depends only on the nature of the counter-electrode (collector) and on the quantum energy, being independent of the nature of the emitting cathode. This conclusion, well known from photoemission measurements at the metal–vacuum interface using the retarding field method, can be regarded as a *sui generis* equivalent of the independence of the threshold potential on the nature of the metal emitter in solution. Here again it should be stressed that both statements are strictly valid for metals with a simple dispersion law, for which the photoelectric and thermionic work functions coincide (cf. Section 4.3).

The analysis of photoelectron emission into solutions includes also a study of the effect of change (in the systems discussed, a decrease) of the work function from its value in vacuum. As was already mentioned in Section 4.4, this change is physically connected with the collective interaction of the electron being emitted with the field of the particles and with the dynamic part of the polarizability of the medium into which emission takes place. A similar change of the work function occurs during electron emission from a metal into dielectrics and semiconductors. Photoemission has become one of the successful methods of investigating surface barriers in diodes, sandwiched layers, and other systems, including metal–semiconductor and metal–dielectric interfaces (cf., e.g., Soshea and Lucas, 1965; Schuermeyer *et al.*, 1968). The work function determined from photoemission measurements can be a few electron volts lower than the work function in vacuum. It has also been indicated that the Fowler formula [Eq. (1.20)] is unsatisfactory in the high-frequency (but still within the near-threshold energy range) region: The photocurrent deviates from Eq. (1.20) toward higher values. Qualitatively, these deviations are in agreement with those predicted by Eq. (1.21) and can be connected with the partial neglect of image forces. Formally, this means that owing to the deviations of the dielectric constant from unity, the condition $\gamma \gtrsim 1$ already obtains in the near-threshold frequency region and the $G(\gamma)$ function increases with γ. It should be mentioned, however, that the observed deviations can, to a considerable extent, be due to other effects, for example, additional generation by light of carriers in a semiconductor or dielectric in which emission takes place.

Another interesting example of systems obeying Eq. (1.21) are metal

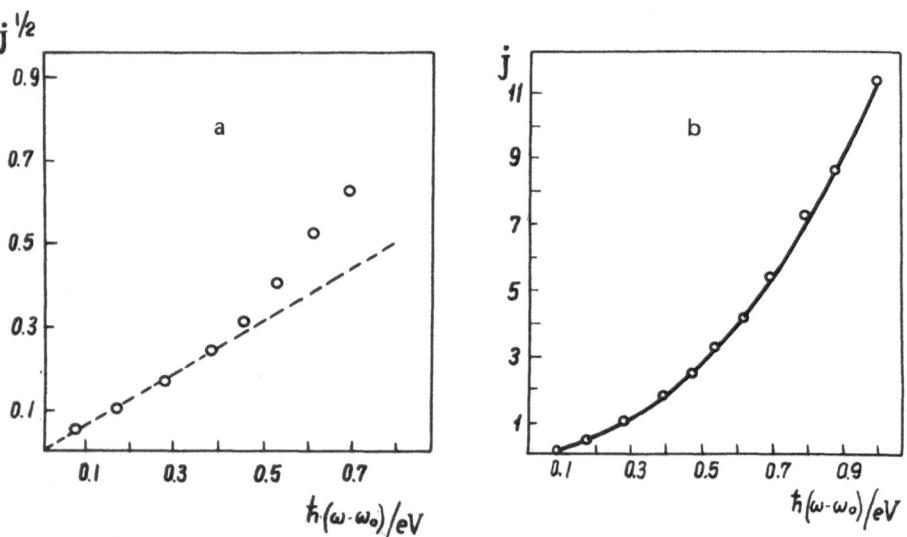

Fig. 10.8. Photoelectron emission from a metal into a dielectric (Barshchevskii and Gurevich, 1970a). (a) The $j^{1/2}$ vs. $\hbar(\omega - \omega_0)$ plot according to the Fowler formula (1.20) for the Ag + AgBr system. (b) The j vs. $\hbar(\omega - \omega_0)$ plot according to Eq. (1.21). $E_e = 0.5$ eV.

particles surrounded by a dielectric. In particular, this is the structure of an exposed but undeveloped layer of a photographic emulsion containing developable centers consisting of silver particles (Galashin and Chibisov, 1968). Illumination of such a layer, placed between two electrodes, by red or infrared light, which by itself does not lead to formation of a latent image, results in the passage of current. The effect is absent in the case of an unexposed emulsion. This indicates the decisive role played by the interaction of light with silver particles formed during the exposure: They emit photoelectrons into the surrounding dielectric, causing the passage of a photocurrent. The dependence of the observed photocurrent on the light frequency (Barshchevskii and Gurevich, 1970a, b) is shown in Fig. 10.8. It can be seen from Fig. 10.8a that the photocurrent j for a Ag + AgBr system obeys Eq. (1.21), whereas Eq. (1.20) for $n = 1$ (according to which $j^{1/2}$ should be a linear function of $\hbar\omega$) shows much poorer agreement with experiment, especially at relatively high frequencies. Comparison of the threshold frequency obtained in this experiment with that characterizing the silver–vacuum interface shows that the decrease of the work function exceeds 2.5 eV.

Results of a similar kind were obtained for the Ag + AgCl system. Thus, the relationships describing the threshold theory of photoemission into dielectrics are in reasonably good agreement with the experimental data. Nevertheless, special experiments aimed at the verification of Eq. (1.21) would certainly be of great interest.

Perspectives of Photoemission Studies

The material presented in this monograph shows that investigations of photoelectron emission into solutions have created a new direction in modern photoelectrochemistry. An attempt is made in the following paragraphs to formulate and summarize the new perspectives opened by this approach.

New Technique

The basic laws of photoelectron emission into solutions were discovered using a relatively simple experimental technique. However, the problems arising often require rather complex apparatus. Using pulsed light sources and measuring devices capable of nanosecond and possibly picosecond resolution, investigations can be undertaken of processes such as the back diffusion of emitted electrons to the electrode, their capture by the metal surface and fast electrochemical reactions of unstable intermediates, etc.

The possibility of high-frequency harmonic light modulation applied for the same purposes by means of crystal-optical shutters seems very tempting.

The successful solution of a number of these problems depends on the feasibility of ultraviolet illumination of photocathodes. Use of a shorter wavelength region than that hitherto applied would enable the electrode potential range studied to be extended toward more positive values so that photoemission methods could then be used to investigate, for example, oxidized metals. The possibilities of using gas-discharge lamps, which have been employed up to now, are, in principle, exhausted. Certain hopes in this direction are connected with such light sources as ultraviolet lasers and synchrotron radiation.

New Objectives

First, the further study of photoemission laws, as well as utilization of photoemission methods, requires extensions of the range of materials under study.

This involves primarily electrode materials with different electronic structures as well as different surface (adsorptive and electrochemical) properties. This field is wide open for further study. Electrodes that have been used up to now have included mercury and a series of solid metals which, in a certain sense, can be called "mercurylike." Their similarity to mercury consists primarily in that they only weakly adsorb hydrogen. Therefore the surface of the emitting electrode is virtually free of adsorbed atoms in the potential range investigated.

However, the study of photoemission at metals covered with adsorbed hydrogen or oxygen layers, which often determine the entire electrochemical behavior of the electrode, is of considerable interest. This pertains primarily to the platinum group metals, which have only recently become the subjects of photoemission studies. The next important class of metals is the iron group. Finally, the photoemission method can supply valuable information concerning relatively thick oxide layers on, for example, valve metals. In this case, the oxide layer can be thought of as modifying, on the one hand, the emissive properties of the support and, on the other hand, as serving itself as the emitter, as in the case of semiconductor materials already investigated.

Second, new solvents which exhibit different types of interaction with excess electrons should be investigated. This may lead to new information concerning thermalization, solvation, and reactivity of electrons in polar and nonpolar liquids.

Finally, the use of various acceptors, both separately and together in the same solution, makes possible measurement of the kinetic characteristics of a wide range of homogeneous reactions, initiated by excess electrons, as well as of electrode reactions including those involving intermediates arising in homogeneous processes. The promising aspects of utilizing photoelectron emission in studies of unstable intermediates require special attention. In this case, the emitting electrode serves simultaneously both as the source of electrons generating the species in the bulk that are to be studied, and as a tool for analyzing the properties of these species. Since the processes investigated occur at distances of the order of 10 Å from the electrode, even relatively short-lived substances can be recorded at the electrode; this ensures a very high resolving ability for the photoemission method.

New Problems

Another interesting direction of research is the study of the chemical behavior of "dry" electrons by means of the photoemission method. Interactions with "dry" electrons become an important process when sufficiently high acceptor concentrations are present in solution.

Photoemission into solutions provides additional possibilities for studying the physical properties of emitting metals and semiconductors (in particular organic semiconductors). Thus, formulation of the possible relationship between the threshold potential and the nature of the metal might involve the properties of the Fermi surface, and thus provide a means of studying these properties. Simultaneous variation of two parameters, light frequency and electrode potential, may lead to a method of investigating the density of electron states near the interface, as well as of collective kinetic effects due, for example, to surface plasmons.

We must finally mention that there is an indirect but important connection between photoemission studies in electrochemical systems and problems of utilization of solar energy, photosynthesis, and, in a broad sense, improved understanding of a variety of light-initiated processes.

Appendixes

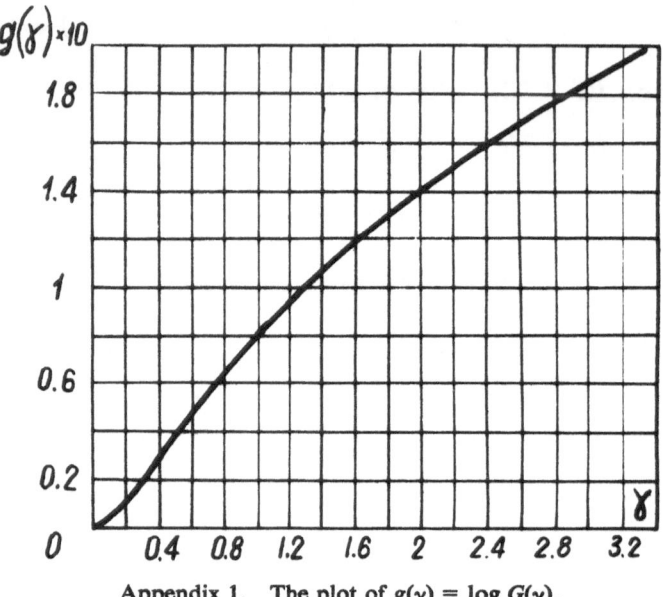

Appendix 1.   The plot of $g(\gamma) \equiv \log G(\gamma)$.

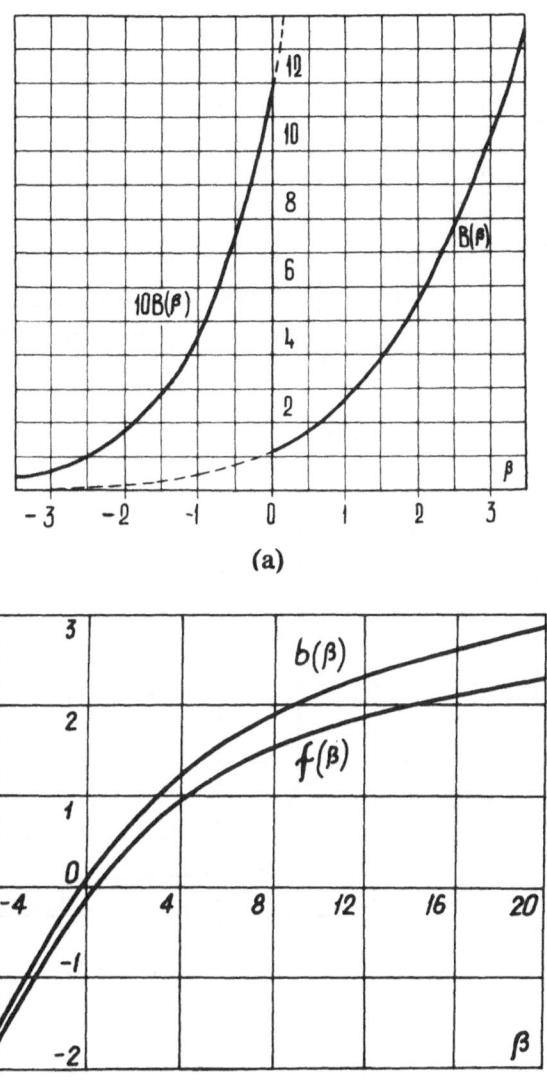

(a)

(b)

Appendix 2.   The plot of (a) $B(\beta)$ and (b) $b(\beta)$ and $f(\beta)$.

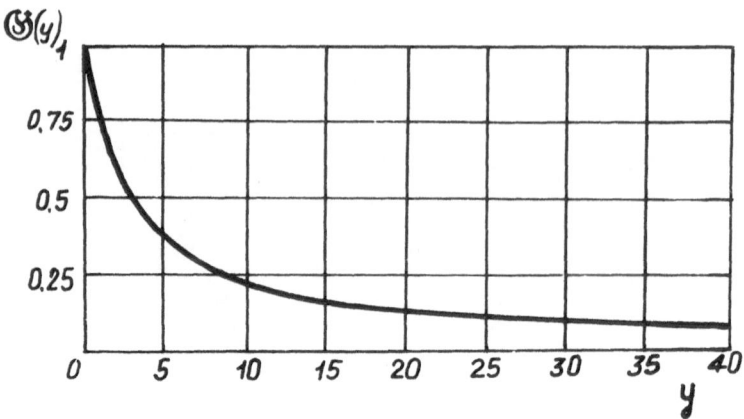

Appendix 3.   The plot of $\mathfrak{G}(y)$.

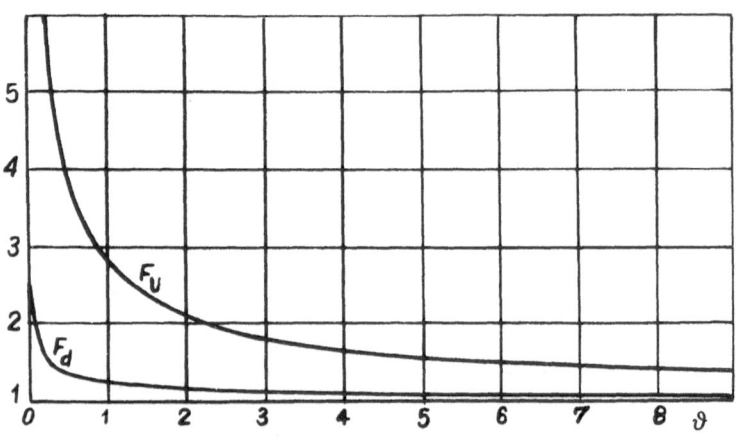

Appendix 4.   The plot of FU and F_d.

Appendix 5. Values of the G (γ) function

γ	$G(\gamma)$	γ	$G(\gamma)$	γ	$G(\gamma)$
0.1	0.007	1.2	1.243	2.3	1.432
0.2	1.026	1.3	1.262	2.4	1.447
0.3	1.049	1.4	1.281	2.5	1.462
0.4	1.073	1.5	1.300	2.6	1.477
0.5	1.096	1.6	1.317	2.7	1.491
0.6	1.119	1.7	1.334	2.8	1.505
0.7	1.141	1.8	1.354	2.9	1.520
0.8	1.162	1.9	1.368	3.0	1.534
0.9	1.183	2.0	1.384	3.1	1.547
1.0	1.204	2.1	1.400		
1.1	1.224	2.2	1.416		

Appendix 6. Values of the $\chi^{0.4}$ function

χ	0	1	2	3	4	5	6	7	8	9	χ
0.00	0.0000	0.0631	0.0633	0.0979	0.1099	0.1201	0.1291	0.1374	0.1450	0.1519	0.00
0.01	0.1585	0.1646	0.1705	0.1760	0.1813	0.1864	0.1913	0.1960	0.2005	0.2049	0.01
0.02	0.2091	0.2132	0.2173	0.2210	0.2249	0.2287	0.2323	0.2358	0.2393	0.2426	0.02
0.03	0.2460	0.2492	0.2524	0.2555	0.2586	0.2616	0.2646	0.2675	0.2703	0.2732	0.03
0.04	0.2759	0.2787	0.2814	0.2840	0.2867	0.2893	0.2918	0.2943	0.2968	0.2993	0.04
0.05	0.3017	0.3041	0.3065	0.3088	0.3111	0.3134	0.3157	0.3179	0.3202	0.3224	0.05
0.06	0.3245	0.3267	0.3288	0.3309	0.3330	0.3351	0.3371	0.3392	0.3412	0.3432	0.06
0.07	0.3452	0.3471	0.3491	0.3510	0.3529	0.3548	0.3567	0.3586	0.3604	0.3623	0.07
0.08	0.3641	0.3659	0.3677	0.3695	0.3713	0.3731	0.3748	0.3765	0.3783	0.3800	0.08
0.09	0.3817	0.3834	0.3850	0.3867	0.3884	0.3900	0.3917	0.3933	0.3949	0.3965	0.09
0.10	0.3961	0.3997	0.4013	0.4028	0.4044	0.4060	0.4075	0.4090	0.4106	0.4120	0.10
0.11	0.4136	0.4151	0.4166	0.4181	0.4195	0.4210	0.4225	0.4239	0.4254	0.4268	0.11
0.12	0.4282	0.4296	0.4311	0.4325	0.4339	0.4353	0.4367	0.4380	0.4394	0.4408	0.12
0.13	0.4422	0.4435	0.4449	0.4462	0.4476	0.4489	0.4502	0.4515	0.4528	0.4542	0.13
0.14	0.4555	0.4568	0.4581	0.4593	0.4606	0.4619	0.4632	0.4644	0.4657	0.4670	0.14
0.15	0.4682	0.4695	0.4707	0.4719	0.4732	0.4744	0.4756	0.4768	0.4780	0.4792	0.15
0.16	0.4804	0.4816	0.4828	0.4840	0.4852	0.4864	0.4876	0.4887	0.4899	0.4911	0.16
0.17	0.4922	0.4934	0.4946	0.4957	0.4968	0.4980	0.4991	0.5003	0.5014	0.5025	0.17
0.18	0.5036	0.5047	0.5059	0.5070	0.5081	0.5092	0.5103	0.5114	0.5125	0.5136	0.18
0.19	0.5146	0.5157	0.5168	0.5179	0.5189	0.5200	0.5211	0.5221	0.5232	0.5243	0.19
0.20	0.5253	0.5264	0.5274	0.5284	0.5295	0.5305	0.5316	0.5326	0.5336	0.5346	0.20
0.21	0.5357	0.5367	0.5377	0.5387	0.5397	0.5407	0.5417	0.5427	0.5437	0.5447	0.21
0.22	0.5457	0.5467	0.5477	0.5487	0.5497	0.5506	0.5516	0.5526	0.5536	0.5545	0.22
0.23	0.5555	0.5565	0.5574	0.5584	0.5594	0.5603	0.5613	0.5622	0.5632	0.5641	0.23
0.24	0.5650	0.5660	0.5669	0.5679	0.5688	0.5697	0.5707	0.5716	0.5725	0.5734	0.24
0.25	0.5743	0.5753	0.5762	0.5771	0.5780	0.5789	0.5798	0.5807	0.5816	0.5825	0.25
0.26	0.5834	0.5843	0.5852	0.5861	0.5870	0.5879	0.5888	0.5897	0.5905	0.5914	0.26
0.27	0.5923	0.5932	0.5941	0.5949	0.5958	0.5967	0.5975	0.5984	0.5993	0.6001	0.27
0.28	0.6010	0.6018	0.6027	0.6036	0.6044	0.6053	0.6061	0.6069	0.6078	0.6086	0.28
0.29	0.6095	0.6103	0.6112	0.6120	0.6128	0.6137	0.6145	0.6153	0.6162	0.6170	0.29
0.0	0.0000	0.1585	0.2091	0.2460	0.2759	0.3017	0.3245	0.3452	0.3641	0.3817	0.0
0.1	0.3981	0.4136	0.4282	0.4422	0.4555	0.4682	0.4804	0.4922	0.5036	0.5146	0.1
0.2	0.5253	0.5357	0.5457	0.5555	0.5650	0.5743	0.5834	0.5923	0.6010	0.6095	0.2
0.3	0.6178	0.6260	0.6340	0.6418	0.6495	0.6571	0.6645	0.6719	0.6791	0.6892	0.3
0.4	0.6931	0.7000	0.7068	0.7135	0.7201	0.7266	0.7330	0.7393	0.7456	0.7518	0.4
0.5	0.7579	0.7639	0.7698	0.7757	0.7816	0.7873	0.7930	0.7986	0.8042	0.8097	0.5
0.6	0.8152	0.8206	0.8260	0.8313	0.8365	0.8417	0.8469	0.8520	0.8570	0.8621	0.6
0.7	0.8670	0.8720	0.8769	0.8817	0.8865	0.8913	0.8960	0.9007	0.9054	0.9100	0.7
0.8	0.9146	0.9192	0.9237	0.9282	0.9326	0.9371	0.9415	0.9458	0.9502	0.9545	0.8
0.9	0.9587	0.9630	0.9672	0.9714	0.9756	0.9797	0.9838	0.9879	0.9920	0.9960	0.9

References

Abramowitz, M., and Stegun, I. A. (1965). *Handbook of Mathematical Functions with Formulas, Graphs and Mathematical Tables*. Dover, New York.

Adams, G. E., and Willson, R. L. (1969). *Trans. Faraday Soc.*, **65**, 2981.

Adawi, I. (1964). *Phys. Rev. Sect. A*, **134**, 788.

Airey, R. L. (1973). *Radiat. Res. Rev.*, **5**, 341.

Aldrich, J. E., Bronskill, M. J., Wolff, R. K., and Hunt, J. W. (1971). *J. Chem. Phys.*, **55**, 530.

Alfaro, V., and Regge, T. (1965). *Potential Scattering*. North Holland, Amsterdam.

Allen, H. C. (1925). *Photoelectricity*. Longmans, Green and Co., London.

Alpatova, N. M., and Grishina, A. D. (1973). *Elektrokhimiya*, **9**, 1375.

Alpatova, N. M., Maltsev, E. I., Vannikov, A. V., and Zabusova, S. E. (1973a). *Elektrokhimiya*, **9**, 1034.

Alpatova, N. M., Fomicheva, M. G., Ovsyannikova, E. V., and Krishtalik, L. I. (1973b). *Elektrokhimiya*, **9**, 1234.

Anand, H. L., and Bhatnagar, S. S. (1928). *Z. Phys. Chem.*, **131**, 134.

Anbar, M., and Neta, P. (1967). *Int. J. Appl. Radiat. Isot.*, **18**, 493.

Ansone, I. K., Rotenberg, Z. A., Slaidin, G. J., and Pleskov, Yu. V. (1976). *Elektrokhimiya*, **12**, 1552.

Antropov, L. I. (1971). In: *Progress in Science. Electrochemistry*, Vol. 6. VINITI, Moscow, p. 5.

Asmus, K. D., and Fendler, J. H. (1968). *J. Phys. Chem.*, **72**, 4285.

Audubert, R. (1923). *Compt. Rend.*, **177**, 818, 1110.

Audubert, R. (1930). *J. Chim. Phys.*, **27**, 169.

Aulich, H., Baron, B., Delahay, P., and Lugo, R. (1973). *J. Chem. Phys.*, **58**, 4439.

Babenko, S. D., Rudenko, T. S., Benderskii, V. A., and Zolotovitskii, Ya. M. (1972). *Fiz. Tverd. Tela*, **14**, 3501.

Babenko, S. D., Benderskii, V. A., and Rudenko, T. S. (1973). *Pis'ma Zh. Eksp. Teor. Fiz.*, **17**, 71.

Babenko, S. D., Benderskii, V. A., Krivenko, A. G., and Rudenko, T. S. (1974a). *Elektrokhimiya*, **10**, 793.

Babenko, S. D., Benderskii, V. A., Krivenko, A. G., and Rudenko, T. S. (1974b). *Fiz. Tverd. Tela*, **16**, 1337.

Bagotskaya, I. A., and Oshe, A. I. (1959). *Proceedings of the Fourth Conference on Electrochemistry*. Izd. Akad. Nauk SSSR, p. 82.

Baksht, F. G., Ivanov, V. G., and Mojzhes, B. Ya. (1971). *Fiz. Tverd. Tela*, **13**, 2896.

Ballard, R. E. (1972). *Chem. Phys. Lett.*, **16**, 300.

Ballard, R. E., and Griffiths, G. A. (1960). *J. Chem. Soc. A*, 1971.

Ballard, R. E., and Griffiths, G. A. (1971). *Chem. Commun.*, 1472.

Barashev, P. P. (1970). *Fiz. Tverd. Tela*, **12**, 1973.

Barker, G. C. (1968). *Electrochim. Acta*, **13**, 1221.

Barker, G. C. (1971). *Ber. Bunsenges. Phys. Chem.*, **75**, 728.

Barker, G. C., and Bolzan, J. A. (1974a). *J. Electroanal. Chem.*, **49**, 227.

Barker, G. C., and Bolzan, J. A. (1974b). *J. Electroanal. Chem.*, **49**, 239.

Barker, G. C., and Bottura, G. (1973a). *J. Electroanal. Chem.*, **46**, 35.

Barker, G. C., and Bottura, G. (1973b). *J. Electroanal. Chem.*, **47**, 199.

Barker, G. C., and Cloke, G. (1974). *J. Electroanal. Chem.*, **52**, 468.

Barker, G. C., and Concialini, V. (1973a). *J. Electroanal. Chem.*, **45**, 320.

Barker, G. C., and Concialini, V. (1973b). *J. Electroanal. Chem.*, **46**, 25.

Barker, G., and Gardner, A. (1965). In: *Fundamental Problems in Contemporary Theoretical Electrochemistry*. Mir, Moscow, p. 118.

Barker, G., and Gardner, A. (1973). *Elektrokhimiya*, **9**, 1684.

Barker, G. C., and Gardner, A. W. (1973). *J. Electroanal. Chem.*, **47**, 205.

Barker, G. C., and McKeown, D. (1975). *J. Electroanal. Chem.*, **62**, 341.

Barker, G. C., Gardner, A. W., and Sammon, D. C. (1966). *J. Electrochem. Soc.*, **113**, 1182.

Barker, G. C., Fowles, P., and Stringer, B. (1970). *Trans. Faraday Soc.*, **66**, 1509.

Barker, G. C., Gardner, A. W., and Bottura, G. (1973). *J. Electroanal. Chem.*, **45**, 21.

Barker, G. C., Bottura, G., Cloke, G., Gardner, A. W., and Williams, M. J. (1974a). *J. Electroanal. Chem.*, **50**, 323.

Barker, G. C., Stringer, B., and Williams, M. J. (1974b). *J. Electroanal. Chem.*, **51**, 305.

Barker, G. C., McKeown, D., Williams, M. J., Bottura, G., and Concialini, V. (1974c). *Faraday Discuss. Chem. Soc.*, No. 56, p. 41.

Baron, B., Chartier, P., Delahay, P., and Lugo, R. (1969). *J. Chem. Phys.*, **51**, 2562.

Baron, B., Delahay, P., and Lugo, R. (1970). *J. Chem. Phys.*, **53**, 1399.

Baron, B., Delahay, P., and Lugo, R. (1971). *J. Chem. Phys.*, **55**, 4180.

Barshchevskii, B. U., and Gurevich, Yu. Ya. (1970a). *Dokl. Akad. Nauk SSSR*, **191**, 115.

Barshchevskii, B. U., and Gurevich, Yu. Ya. (1970b). *Fiz. Tverd. Tela*, **12**, 3380.

Baumann, F. (1960). *Z. Phys.*, **158**, 607.

Baz', A. I., Zel'dovich, Ya. B., and Perelomov, A. M. (1966). *Scattering, Reactions, and Disintegrations in Nonrelativistic Quantum Mechanics*. Nauka, Moscow.

Becquerel, E. (1839). *Compt. Rend.*, **9**, 145.

Becquerel, E. (1859). *Ann. Chim. Phys.*, **56**, 99.

Bell, R. L., and Spicer, W. E. (1970). *Proc. IEEE*, **58**, 1788.

Benderskii, V. A., Babenko, S. D., Zolotovitskii, Ya. M., Krivenko, A. G., and Rudenko, T. S. (1974). *J. Electroanal. Chem.*, **56**, 325.

Berg, H. (1960a). *Naturwissenschaften*, **47**, 320.

Berg, H. (1960b). *Coll. Czech. Chem. Commun.*, **25**, 3404.

Berg, H. (1961). *Ber. Bunsenges. Phys. Chem.*, **65**, 710.

Berg, H. (1966). In: *Modern Aspects of Polarography*. T. Kambara, ed. Plenum Press, New York, p. 29.

Berg, H. (1968). *Electrochim. Acta*, **13**, 1249.

Berg, H., and Reissmann, P. (1970). *J. Electroanal. Chem.*, **24**, 427.

Berg, H., and Schweiss, H. (1965). In: *Fundamental Problems in Contemporary Theoretical Electrochemistry*. Mir, Moscow, p. 130.

Berg, H., Schweiss, H., Stutter, E., and Weller, K. (1967). *J. Electroanal. Chem.*, **15**, 415.

Bethe, H. A., and Salpeter, E. E. (1957). *Quantum Mechanics of One- and Two-Electron Atoms*. Springer, Berlin–Göttingen–Heidelberg.

Bewick, A., Conway, B. E., and Tuxford, A. M. (1973). *J. Electroanal. Chem.*, **42**, Appendix 11.

Boikova, G. V., Krotova, M. D., and Pleskov, Yu. V. (1976). *Elektrokhimiya*, **12**, 922.

Bomchil, G., Schiffrin, D. J., and D'Alessio, J. T. (1970). *J. Electroanal. Chem.*, **25**, 107.

Borbat, A. M., Gorban', I. S., Okhrimenko, B. A., Subbota-Melnik, P. A., Shaikevich, I. A., and Shishlovskii, A. A. (1967). *Optical Measurements*. Tekhnika, Kiev.

Born, M., and Wolf, E. (1964). *Principles of Optics*. Pergamon, Oxford–London–Edinburgh–New York–Paris–Frankfurt.

Bottura, G., Bubani, B., and Barker, G. C. (1975). *J. Electroanal. Chem.*, **62**, 259.

Bowden, F. P. (1931). *Trans. Faraday Soc.*, **27**, 505.

Brauer, M. (1966). *Phys. Status Solidi*, **14**, 413.

Brodskii, A. M., and Frumkin, A. N. (1970). *Elektrokhimiya*, **6**, 658.

Brodskii, A. M., and Gurevich, Yu. Ya. (1973). *Theory of Electronic Emission from Metals*. Nauka, Moscow.

Brodskii, A. M., and Tsarevskii, A. V. (1973). *Elektrokhimiya*, **9**, 1671.

Brodskii, A. M., and Tsarevskii, A. V. (1974). *Elektrokhimiya*, **10**, 1635.

Brodskii, A. M., and Tsarevskii, A. V. (1975a). *Fifth All-Union Conference on Electrochemistry*. Abstracts of Proceedings. Moscow, p. 130.

Brodskii, A. M., and Tsarevskii, A. V. (1975b). *Zh. Eksp. Teor. Fiz.*, **69**, 936.

Brodskii, A. M., Gurevich, Yu. Ya., and Sheberstov, S. V. (1971). *J. Electroanal. Chem.*, **32**, 353.

Bronskill, M. J., Wolff, R. K., and Hunt, J. W. (1970). *J. Chem. Phys.*, **53**, 4201, 4211.

Brooks, J. M., and Dewald, R. R. (1968). *J. Phys. Chem.*, **72**, 2655.

Bunkin, F. V., and Fedorov, M. V. (1965). *Zh. Eksp. Teor. Fiz.*, **48**, 1341.

Caldin, E. F. (1964). *Fast Reactions in Solution*. Blackwell, Oxford.

Cardona, M. (1969). *Modulation Spectroscopy*. Academic Press, New York–London.

Case, B., and Parsons, R. (1967). *Trans. Faraday Soc.*, **63**, 1224.

Clark, P. E., and Garrett, A. B. (1939). *J. Am. Chem. Soc.*, **61**, 1805.

Concialini, V., Tubertini, O., and Barker, G. C. (1974). *J. Electroanal. Chem.*, **57**, 413.

Conway, B. E. (1972). In: *Modern Aspects of Electrochemistry*, Vol. 7. Plenum Press, New York, p. 83.

Conway, B. E., and MacKinnon, D. J. (1970). *J. Phys. Chem.*, **74**, 3663.

Copeland, A. W., Black, O. D., and Garrett, A. B. (1942). *Chem. Rev.*, **31**, 177.

Copeland, D. A., Kestner, N. B., and Jortner, J. (1970). *J. Chem. Phys.*, **53**, 1189.

Crowell, J., and Ritchie, R. H. (1970). *J. Opt. Soc. Am.*, **60**, 794.

Damaskin, B. B., and Kaganovich, R. I. (1977). *Elektrokhimiya* (in press).

Damaskin, B. B., and Nikolaeva-Fedorovich, N. V. (1961). *Zh. Fiz. Khim.*, **35**, 1279.

Damaskin, B. B., Survila, A. A., and Rybalka, L. E. (1967). *Elektrokhimiya*, **3**, 146.

Davydov, A. S. (1948). *Zh. Eksp. Teor. Fiz.*, **18**, 913.

Davison, S. G., and Levine, J. D. (1970). *Surface States*. Academic Press, New York–London.

Dejgen, M. F. (1954). *Tr. Inst. Fiz. Akad. Nauk Ukr. SSR*, **5**, 119.

Delahay, P. (1965). *Double Layer and Electrode Kinetics*. Wiley-Interscience, New York–London–Sydney.

Delahay, P. (1971). *J. Chem. Phys.*, **55**, 4188.

Delahay, P. (1973). In: *Electrons in Fluids. The Nature of Metal-Ammonia Solutions*. J. Jortner and N. R. Kestner, eds. Springer, Berlin–Heidelberg–New York, p. 131.

Delahay, P., and Srinivasan, V. S. (1966). *J. Phys. Chem.*, **70**, 420.

Delahay, P., Chartier, P., and Nemec, L. (1970). *J. Chem. Phys.*, **53**, 3126.

De Levie, R., and Husovsky, A. A. (1969). *J. Electroanal. Chem.*, **20**, 181.

De Levie, R., and Kreuser, J. C. (1969). *J. Electroanal. Chem.*, **21**, 221.

Dobretsov, L. N., and Gomoyunova, M. V. (1966). *Emission Electronics*. Nauka, Moscow.

Dogonadze, R. R., Krishtalik, L. I., and Pleskov, Yu. V. (1974). *Elektrokhimiya*, **10**, 507.

Dolin, P. I., and Ershler, B. V. (1962). *Proceedings of the Second All-Union Conference on Radiation Chemistry*. Izd. Akad. Nauk SSSR, Moscow, p. 87.

Dorfman, L. M. (1965). In: *Solvated Electron*. Am. Chem. Soc. Publ., Washington, p. 36.

Eletskii, V. V., and Pleskov, Yu. V. (1974). *Elektrokhimiya*, **10**, 179.

Eletskii, V. V., Rotenberg, Z. A., and Pleskov, Yu. V. (1969). *Elektrokhimiya*, **5**, 469.

Eletskii, V. V., Rotenberg, Z. A., and Pleskov, Yu. V. (1970). *Elektrokhimiya*, **6**, 1244.

Eletskii, V. V., Rotenberg, Z. A., and Pleskov, Yu. V. (1971). *Khim. Vys. Energ.*, **5**, 325.

Endriz, J. G., and Spicer, W. E. (1971). *Phys. Rev. Lett.*, **27**, 570.

Erenburg, R. G., Krishtalik, L. I., and Yaroshevskaya, I. P. (1975). *Elektrokhimiya*, **11**, 1072, 1076, 1244.

Ershler, B. V. (1952). *Usp. Khim.*, **21**, 237.

Farkas, G., Naray, Z., and Varga, P. (1967). *Phys. Lett. A*, **24**, 475.

Feibelman, P. J. (1974). *Surf. Sci.*, **46**, 558.

Feibelman, P. J., and Eastman, D. E. (1974). *Phys. Rev.*, **10**, 4932.

Fetter, K. J. (1961). *Elektrochemische Kinetik*. Springer, Berlin–Göttingen–Heidelberg.

Fischer, T. E. (1972). *J. Vac. Sci. Technol.*, **9**, 860.

Fomenko, V. S. (1970). *Emission Properties of Materials*. Naukova Dumka, Kiev.

Fowler, R. H. (1931). *Phys. Rev.*, **38**, 45.

Frumkin, A. N. (1933). *Z. Phys. Chem.*, **164**, 121.

Frumkin, A. N. (1935). Quoted in: Horiuti, J., and Polanyi, M., *Acta Physicochim. URSS*, **2**, 505.

Frumkin, A. N. (1957). *Zh. Fiz. Khim.*, **31**, 1875.

Frumkin, A. N. (1961). In: *Advances in Electrochemistry and Electrochemical Engineering*, Vol. 1. P. Delahay, ed. Interscience, New York, p. 65.

Frumkin, A. N. (1965a). *Elektrokhimiya*, **1**, 394.

Frumkin, A. N. (1965b). *Sven. Kem. Tidsk.*, **77**, 300.

Frumkin, A. N. (1972). In: *Double Layer and Adsorption on Solid Electrodes*. II. Tartu, p. 5.

Frumkin, A. N., and Damaskin, B. B. (1959). *Dokl. Akad. Nauk SSSR*, **129**, 862.

Frumkin, A. N., and Damaskin, B. B. (1975). *Dokl. Akad. Nauk SSSR*, **221**, 395.

Frumkin, A. N., Bagotskii, V. S., and Jofa, Z. A. (1951). *Zh. Fiz. Khim.*, **25**, 1117.

Frumkin, A. N., Bagotskii, V. S., Jofa, Z. A., and Kabanov, B. N. (1952). *Kinetics of Electrode Processes*. Izd. Mosk. Univ., Moscow.

Frumkin, A. N., Jofa, Z. A., and Gerovich, M. A. (1956). *Zh. Fiz. Khim.*, **30**, 1455.

Frumkin, A. N., Petry, O. A., and Nikolaeva-Fedorovich, N. V. (1959). *Dokl. Akad. Nauk SSSR*, **128**, 1006.

Frumkin, A. N., Petry, O. A., and Nikolaeva-Fedorovich, N. V. (1962). *Dokl. Akad. Nauk SSSR*, **147**, 878.

Gadzuk, J. W. (1970). *Phys. Rev. B*, **1**, 1267.

Galashin, E. A., and Chibisov, K. V. (1968). *Dokl. Akad. Nauk SSSR*, **178**, 872.

Ganzhina, I. M., Damaskin, B. B., Kaganovich, R. I., and Ivanova, R. V. (1971). *Elektrokhimiya*, **7**, 362.

Ganzhina, I. M., Damaskin, B. B., and Kaganovich, R. I. (1972). *Elektrokhimiya*, **8**, 93.

Gerischer, H. (1966a). *J. Electrochem. Soc.*, **113**, 1174.

Gerischer, H. (1966b). *J. Electrochem. Soc.*, **113**, 1199.

Gerischer, H. (1973). *Ber. Bunsenges. Phys. Chem.*, **77**, 771.

Gerischer, H., Mayer, E., and Sass, J. K. (1972). *Ber. Bunsenges. Phys. Chem.*, **76**, 1191.

Gokhshtein, A. Ya., and Gokhshtein, Ya. A. (1962). *Zh. Fiz. Khim.*, **36**, 651.

Goldberger, M. L., and Watson, K. M. (1964). *Collision Theory*. Wiley, New York–London–Sydney.

Gorodetskaya, A., and Frumkin, A. (1938). *Dokl. Akad. Nauk SSSR*, **18**, 649.

Görlich, P. (1962). *Photoeffekte*, Bd. 1. Akademische Verlag-gesellschaft, Leipzig.

Grahame, D. C. (1947). *Chem. Rev.*, **41**, 441.

Grahame, D. C. (1957). *J. Am. Chem. Soc.*, **79**, 2093.

Gremmo, N., and Randles, J. E. B. (1974). *J. Chem. Soc., Faraday Trans. 1*, 1488.

Grigoryev, N. B., Gedvillo, I., and Bardina, N. G. (1972). *Elektrokhimiya*, **8**, 409.

Gurevich, Yu. Ya. (1969). *Fiz. Tverd. Tela*, **11**, 2976.

Gurevich, Yu. Ya. (1972). *Elektrokhimiya*, **8**, 1564.

Gurevich, Yu. Ya. (1974). In: *Electron Processes on Semiconductor Surfaces*. Novosibirsk, p. 93.

Gurevich, Yu. Ya., and Rotenberg, Z. A. (1968). *Elektrokhimiya*, **4**, 529.

Gurevich, Yu. Ya., Brodskii, A. M., and Levich, V. G. (1967). *Elektrokhimiya*, **3**, 1302.

Hart, E. J., and Anbar, M. (1970). *The Hydrated Electron*. Wiley-Interscience, New York–London–Sydney–Toronto.

Häsing, J. (1940). *Ann. Phys.*, 5. Folge, **37**, 509.

Hertz, H. (1887). *Ann. Phys.*, **31**, 983.

Heyrovsky, J., and Kuta, J. (1962). *Zaklady polarografie*. Naklad CSAV, Praha.

Heyrovsky, M. (1965). *Nature*, **200**, 1356.

Heyrovsky, M. (1966a). Thesis. Cambridge University.

Heyrovsky, M. (1966b). *Nature*, **209**, 708.

Heyrovsky, M. (1967a). *Z. Phys. Chem.*, N.F., **52**, 1.

Heyrovsky, M. (1967b). *Proc. R. Soc. (London) Ser. A*, **301**, 411.

Heyrovsky, M. (1973). *Croat. Chem. Acta*, **45**, 247.

Hillson, P. J., and Rideal, E. K. (1949). *Proc. R. Soc. (London) Ser. A*, **199**, 295.

Hillson, P. J., and Rideal, E. K. (1953). *Proc. R. Soc. (London) Ser. A*, **216**, 458.

Hoffmann, J., and Steinmann, N. (1968). *Phys. Status Solidi*, **30**, K53.

Holub, K. (1969). *J. Electroanal. Chem.*, **23**, Appendix 13.

Honda, K. (1969). *Kogyo Kagaku Zasshi*, **72**, 63.

Horiuti, J. (1958). *Z. Phys. Chem.*, N.F., **15**, 162.

Hughes, A. L., and Du Bridge, L. A. (1932). *Photoelectric Phenomena*. McGraw-Hill, New York–London.

Imai, H. (1973). In: *Electrochemistry in Non-Aqueous Solvents and Its Application in Analytical Chemistry* (U.S.–Japan Cooperative Science Seminar). Tokyo, p. 5.

Itskovich, F. I. (1966). *Zh. Eksp. Teor. Fiz.*, **51**, 301.

Itskovich, F. I. (1967). *Zh. Eksp. Teor. Fiz.*, **52**, 1720.

James, L. W., and Moll, J. L. (1969). *Phys. Rev.*, **183**, 740.

Jofa, Z. A., and Mikulin, K. P. (1944). *Zh. Fiz. Khim.*, **18**, 137.

Jonah, C. D., Hart, E. J., and Matheson, M. S. (1973). *J. Phys. Chem.*, **77**, 1838.

Jortner, J. (1959). *J. Chem. Phys.*, **30**, 839.

Jortner, J., and Kestner, N. R. (eds.) (1973). *Electrons in Fluids. The Nature of Metal–Ammonia Solutions.* Springer, Berlin–Heidelberg–New York.

Jost, R. (1947). *Helv. Phys. Acta*, **20**, 256.

Kabanov, B. N. (1936). *Zh. Fiz. Khim.*, **8**, 486.

Kajiwara, T., Funabashi, K., and Naleway, C. (1972). *Phys. Rev. Sect. A*, **6**, 808.

Kane, E. O. (1962). *Phys. Rev.*, **127**, 131.

Kanevskii, E. (1950). *Zh. Fiz. Khim.*, **24**, 1511.

Karapetyanz, M. Kh., and Karapetyanz, M. L. (1968). *Main Thermodynamic Constants of Inorganic Substances.* Khimiya, Moscow.

Kenney, G. A., and Walker, D. C. (1971). In: *Electroanalytical Chemistry*, Vol. 5, A. J. Bard, ed. Marcel Dekker, New York, p. 1.

Koller, L. R. (1965). *Ultraviolet Radiation.* Wiley, New York–London–Sydney.

Korn, G. A., and Korn, T. M. (1961). *Mathematical Handbook for Scientists and Engineers.* McGraw-Hill, New York–Toronto–London.

Korndorf, S. F., Dubinovskii, A. M., Muromova, N. S., Perova, N. I., and Surova, E. Ya. (1967). *Calculation of Photoelectric Chains.* Energiya, Moscow.

Korshunov, L. I., Benderskii, V. A., and Zolotovitskii, Ya. M. (1968a). *Elektrokhimiya*, **4**, 499.

Korshunov, L. I., Benderskii, V. A., Gol'danskii, V. I., and Zolotovitskii, Ya. M. (1968b). *Pis'ma Zh. Eksp. Teor. Fiz.*, **7**, 55.

Korshunov, L. I., Zolotovitskii, Ya. M., and Benderskii, V. A. (1969). *Elektrokhimiya*, **5**, 716.

Korshunov, L. I., Zolotovitskii, Ya. M., Benderskii, V. A., and Gol'danskii, V. I. (1970a). *Khim. Vys. Energ.*, **4**, 346.

Korshunov, L. I., Zolotovitskii, Ya. M., Benderskii, V. A., and Gol'danskii, V. I. (1970b). *Khim. Vys. Energ.*, **4**, 461.

Korshunov, L. I., Zolotovitskii, Ya. M., and Benderskii, V. A. (1971). *Usp. Khim.*, **40**, 1511.

Kraus, C. A. (1921). *J. Am. Chem. Soc.*, **43**, 749.

Krishtalik, L. I. (1965). *Usp. Khim.*, **34**, 1831.

Krishtalik, L. I. (1968). *Elektrokhimiya*, **4**, 877.

Krishtalik, L. I. (1969). *Elektrokhimiya*, **5**, 3.

Krishtalik, L. I. (1972). *J. Electroanal. Chem.*, **35**, 157.

Krishtalik, L. I., and Alpatova, N. M. (1975). *J. Electroanal. Chem.*, **65**, 219.

Krishtalik, L. I., and Alpatova, N. M. (1976). *Elektrokhimiya*, **12**, 163.

Krotova, M. D., and Pleskov, Yu. V. (1973). *Fiz. Tverd. Tela*, **15**, 2806.

Kryukova, T. A. (1949). *Dokl. Akad. Nauk SSSR*, **65**, 517.

Laar, J., and Scheer, J. (1968). *Phillips Tech. Rev.*, **29**, 54.

Lam, K. Y., and Hunt, J. W. (1975). *Int. J. Radiat. Phys. Chem.*, **7**, 317.

Landau, L. D., and Lifshitz, E. M. (1959). *Electrodynamics of Continuous Media.* Fizmatgiz, Moscow.

Landau, L. D., and Lifshitz, E. M. (1974). *Quantum Mechanics. Nonrelativistic Theory.* Nauka, Moscow.

Langreth, D. C. (1971). *Phys. Rev. Sect. B*, **3**, 3120.

Leikis, D. I., Rybalka, K. V., Sevastyanov, E. S., and Frumkin, A. N. (1973). *J. Electroanal. Chem.*, **46**, 161.

Levin, I., and Delahay, P. (1970). *J. Electroanal. Chem.*, **24**, Appendix 17.

Levina, S. D., and Kalish, T. V. (1956). *Dokl. Akad. Nauk SSSR*, **109**, 97.

Levina, S., and Zarinskii, V. (1937). *Acta Physicochim. URSS*, **7**, 485.

Lenard, P. (1900). *Ann. Phys.*, **2**, 359.
Litvak, V. L. (1966). *Photoelectric Sensors in Control, Guidance, and Regulating Systems.* Nauka, Moscow.
Lugo, R., and Delahay, P. (1972). *J. Chem. Phys.*, **57**, 2122.
Mahan, G. D. (1970). *Phys. Rev. Sect. B*, **2**, 4334.
Malev, V. V. (1970). *Elektrokhimiya*, **6**, 676, 862, 1917.
Many, A., Goldstein, A., and Grover, N. B. (1965). *Semiconductor Surfaces.* North Holland, Amsterdam.
Marshak, I. S. (1963). *Pulsed Light Sources.* Gosenergoizdat, Moscow–Leningrad.
Matsuda, H., and Delahay, P. (1960). *J. Phys. Chem.*, **64**, 332.
Matsuda, A., and Horiuti, J. (1958). *J. Res. Inst. Catal., Hokkaido Univ.*, **6**, 231.
Matthews, D. B., and Khan, S. U. M. (1975). *Aust. J. Chem.*, **28**, 253.
Means, D. K., and Mark, H. B. (1972). *Chem. Instrum.*, **3**, 271.
Meessen, A. (1968). *Phys. Status Solidi*, **26**, 125.
Mehl, W., and Hale, J. M. (1967). In: *Advances in Electrochemistry and Electrochemical Engineering*, Vol. 6, P. Delahay, ed. Interscience, New York, p. 399.
Meyer, E. (1973). Dissertation, Tech. Univ. München.
Mott, N. F., and Davis, E. A. (1971). *Electronic Processes in Noncrystalline Materials.* Clarendon Press, Oxford.
Mozumder, A., and Magee, J. L. (1966). *Radiat. Res.*, **28**, 203.
Mozumder, A., and Magee, J. L. (1967). *J. Chem. Phys.*, **47**, 939.
Mozumder, A., and Magee, J. L. (1975). *Int. J. Radiat. Phys. Chem.*, **7**, 83.
Mustel', E. R., and Parygin, V. N. (1970). *Methods of Light Modulation and Scanning.* Nauka, Moscow.
Myamlin, V. A., and Pleskov, Yu. V. (1967). *Electrochemistry of Semiconductors.* Plenum Press, New York.
Nemec, L. (1973). *J. Chem. Phys.*, **59**, 6092.
Nemec, L., and Delahay, P. (1972). *J. Chem. Phys.*, **57**, 2135.
Nemec, L., Baron, B., and Delahay, P. (1972). *Chem. Phys. Lett.*, **16**, 278.
Newton, R. G. (1967). *Scattering Theory of Waves and Particles.* McGraw-Hill, New York–Toronto–London–Sydney.
Nikolaeva-Fedorovich, N. V., Frumkin, A. N., and Keis, H. A. (1970). In: *Double Layer and Adsorption on Solid Electrodes.* II. Tartu, p. 262.
Palm, U., and Tenno, T. (1973). *J. Electroanal. Chem.*, **42**, 457.
Parker, C. A. (1968). *Photoluminescence of Solutions with Applications to Photochemistry and Analytical Chemistry.* Elsevier, Amsterdam.
Parsons, R. (1954). In: *Modern Aspects of Electrochemistry*, J. O'M. Bockris and B. E. Conway, eds. Butterworths, London.
Parsons, R. (1964). *Surf. Sci.*, **2**, 418.
Petrov, N. N. (1971). *Zh. Tech. Fiz.*, **61**, 2473.
Petry, O. A., and Nikolaeva-Fedorovich, N. V. (1961). *Dokl. Akad. Nauk SSSR*, **141**, 1139.
Philips, J. C. (1966). In: *The Optical Properties of Solids.* Academic Press, New York–London, p. 323.
Pikaev, A. K. (1969). *The Solvating Electron in Radiation Chemistry.* Nauka, Moscow.
Pikaev, A. K. (1970). *Vestn. Akad. Nauk SSSR*, No. 8, p. 97.
Pines, D. (1963). *Elementary Excitations in Solids.* Benjamin, New York.
Platzman, R. L. (1953). In: *Physical and Chemical Aspects of Basic Mechanisms in Radiobiology.* J. L. Magee *et al.*, eds. U.S. Natl. Acad. Sci. Publ. No. 305, Washington, p. 34.

Pleskov, V. A., and Ershler, B. V. (1949). *Zh. Fiz. Khim.*, **23**, 101.

Pleskov, Yu. V., and Filinovskii, V. Yu. (1976). *Rotating Disc Electrode*. Plenum Press, New York.

Pleskov, Yu. V., and Rotenberg, Z. A. (1969). *J. Electroanal. Chem.*, **20**, 1.

Pleskov, Yu. V., and Rotenberg, Z. A. (1974). *Khim. Vys. Energ.*, **8**, 99.

Pleskov, Yu. V., Rotenberg, Z. A., and Lakomov, V. I. (1970). *Elektrokhimiya*, **6**, 1787.

Pleskov, Yu. V., Rotenberg, Z. A., Eletskii, V. V., and Lakomov, V. I. (1974). *Faraday Discuss. Chem. Soc.*, No. 56, p. 52.

Postl, D., and Schindewolf, U. (1971). *Ber. Bunsenges. Phys. Chem.*, **75**, 662.

Prishchepa, Yu. A., Rotenberg, Z. A., and Pleskov, Yu. V. (1975). *J. Electroanal. Chem.*, **66**, 3.

Punning, K., and Past, V. (1969). *Uch. Zap. Tartusk. Gos. Univ.*, No. 235, p. 35.

Pyle, T., and Roberts, C. (1968). *J. Electrochem. Soc.*, **115**, 247.

Randles, J. E. B. (1956). *Trans. Faraday Soc.*, **52**, 1573.

Rangarajan, S. K. (1963). *J. Electroanal. Chem.*, **5**, 350.

Ritchie, R. H. (1957). *Phys. Rev.*, **106**, 874.

Ritchie, R. H., and Wilems, R. E. (1969). *Phys. Rev.*, **178**, 372.

Rivière, J. C. (1969). In: *Solid State Surface Science*, Vol. 1, Mino Green, ed. Marcel Dekker, New York.

Rokhlin, G. N. (1966). *Gas Discharge Light Sources*. Energiya, Moscow–Leningrad.

Rotenberg, Z. A. (1972). *Elektrokhimiya*, **8**, 1198.

Rotenberg, Z. A. (1973). *Elektrokhimiya*, **9**, 511.

Rotenberg, Z. A. (1974). *Elektrokhimiya*, **10**, 1031.

Rotenberg, Z. A., and Gurevich, Yu. Ya. (1968). *Elektrokhimiya*, **4**, 984.

Rotenberg, Z. A., and Gurevich, Yu. Ya. (1973). *Elektrokhimiya*, **9**, 159.

Rotenberg, Z. A., and Gurevich, Yu. Ya. (1975). *J. Electroanal. Chem.*, **66**, 165.

Rotenberg, Z. A., and Pleskov, Yu. V. (1968). *Elektrokhimiya*, **4**, 826.

Rotenberg, Z. A., and Pleskov, Yu. V. (1969). *Elektrokhimiya*, **5**, 982.

Rotenberg, Z. A., and Pleskov, Yu. V. (1970). *Elektrokhimiya*, **6**, 418.

Rotenberg, Z. A., and Pleskov, Yu. V. (1973). *Elektrokhimiya*, **9**, 1419.

Rotenberg, Z. A., Pleskov, Yu. V., and Lakomov, V. I. (1968a). *Elektrokhimiya*, **4**, 1022.

Rotenberg, Z. A., Gurevich, Yu. Ya., and Pleskov, Yu. V. (1968b). *Elektrokhimiya*, **4**, 1086.

Rotenberg, Z. A., Lakomov, V. I., and Pleskov, Yu. V. (1970a). *J. Electroanal. Chem.*, **27**, 403.

Rotenberg, Z. A., Lakomov, V. I., Brodskii, A. M., and Pleskov, Yu. V. (1970b). *Elektrokhimiya*, **6**, 1387.

Rotenberg, Z. A., Lakomov, V. I., and Pleskov, Yu. V. (1972). *Elektrokhimiya*, **8**, 313.

Rotenberg, Z. A., Lakomov, V. I., and Pleskov, Yu. V. (1973a). *Elektrokhimiya*, **9**, 11.

Rotenberg, Z. A., Lakomov, V. I., and Pleskov, Yu. V. (1973b). *Elektrokhimiya*, **9**, 152.

Rotenberg, Z. A., Prishchepa, Yu. A., and Pleskov, Yu. V. (1974). *J. Electroanal. Chem.*, **56**, 345.

Rotenberg, Z. A., Prishchepa, Yu. A., and Ansone, I. K. (1975a). *Elektrokhimiya*, **11**, 651.

Rotenberg, Z. A., Prishchepa, Yu. A., Ansone, I. K., Slaidin, G. J., and Pleskov, Yu. V. (1975b). In: *Double Layer and Adsorption on Solid Electrodes*. IV. Tartu, p. 264.

Sass, J. K., Sen, R. K., Meyer, E., and Gerischer, H. (1974). *Surf. Sci.*, **44**, 515.
Schaich, W. L., and Ashcroft, N. W. (1970). *Solid State Commun.*, **8**, 1959.
Schaich, W. L., and Ashcroft, N. W. (1971). *Phys. Rev. Sect. B*, **3**, 2452.
Schiffrin, D. J. (1972). *Croat. Chem. Acta*, **44**, 139.
Schiffrin, D. J. (1974). *Faraday Discuss. Chem. Soc.*, **56**, 41.
Schuermeyer, F. L., Young, R., and Blasingame, J. M. (1968). *J. Appl. Phys.*, **39**, 1791.
Sharma, V. P., Delahay, P., Susbielles, G. G., and Tessari, G. (1968). *J. Electroanal. Chem.*, **16**, 285.
Sheberstov, S. V., Brodskii, A. M., and Gurevich, Yu. Ya. (1970), *Elektrokhimiya*, **6**, 1182.
Shurkliff, W. A. (1962). *Polarized Light*. Harvard University Press, Cambridge.
Sihvonen, V. (1926). *Ann. Acad. Sci. Fenn. Ser. A*, **26**, 3.
Soboleva, N. A. (1973). *Usp. Fiz. Nauk*, **111**, 331.
Soshea, R. W., and Lucas, R. C. (1965). *Phys. Rev. Sect. A*, **138**, 1182.
Spicer, W. E. (1968). In: *A Survey of Phenomena in Ionized Gases*. International Atomic Energy Agency, Vienna, p. 271.
Stern, E. A., and Ferrell, R. A. (1960). *Phys. Rev.*, **120**, 130.
Stoletov, A. (1888a). *Compt. Rend.*, **106**, 1149.
Stoletov, A. (1888b). *Compt. Rend.*, **106**, 1593.
Tamm, I. E., and Shubin, S. P. (1931). *Z. Phys.*, **68**, 97.
Teal, G. K. (1947). *Phys. Rev.*, **71**, 138.
Temkin, M. I. (1953). In: *Proceedings of a Conference on Electrochemistry*. Izd. Akad. Nauk SSSR, Moscow, p. 181.
Temkin, M. I., and Frumkin, A. N. (1955). *Zh. Fiz. Khim.*, **29**, 1513.
Thomson, J. J. (1899). *Philos. Mag.*, **48**, 577.
Trasatti, S. (1971). *J. Electroanal. Chem.*, **33**, 351.
Trasatti, S. (1974). *J. Electroanal. Chem.*, **52**, 313.
Veselovskii, V. I. (1946). *Zh. Fiz. Khim.*, **20**, 1493.
Vinnikov, Yu. Ya., Shepelin, V. A., and Veselovskii, V. I. (1973). *Elektrokhimiya*, **9**, 1557.
Vinnikov, Yu. Ya., Shepelin, V. A., and Veselovskii, V. I. (1974). *Elektrokhimiya*, **10**, 650.
Walker, D. C. (1967a). *Can. J. Chem.*, **45**, 807.
Walker, D. C. (1967b). *Q. Rev. (London)*, **21**, 79.
Willardson, R. K., and Beer, A. C. (eds.) (1967). *Optical Properties of III-V Compounds*. Academic Press, New York–London.
Yamashita, K., and Imai, H. (1975). *Denki Kagaku*, **43**, 386.
Young, L. (1961). *Anodic Oxide Films*. Academic Press, New York–London.
Zaidel', A. N., and Shreider, E. Ya. (1967). *Spectroscopy of Vacuum Ultraviolet*. Nauka, Moscow.
Ziman, J. M. (1960). *Electrons and Phonons*. Clarendon Press, Oxford.
Ziman, J. M. (1972). *Principles of the Theory of Solids*. Cambridge, University Press.
Zolotovitskii, Ya. M., Korshunov, L. I., Benderskii, V. A., and Bartenev, V. Ya. (1971). *Izv. Akad. Nauk SSSR, Ser. Khim.*, 1444.
Zolotovitskii, Ya. M., Korshunov, L. I., and Benderskii, V. A. (1972a). *Izv. Akad. Nauk SSSR, Ser. Khim.*, 802.
Zolotovitskii, Ya. M., Benderskii, V. A., Babenko, S. A., Korshunov, L. I., and Rudenko, T. S. (1972b). In: *Double Layer and Adsorption on Solid Electrodes*. III. Tartu, p. 119.
Zolotovitskii, Ya. M., Benderskii, V. A., Babenko, S. D., and Krivenko, A. T. (1975). In: *Double Layer and Adsorption on Solid Electrodes*. IV. Tartu, p. 100.